Air Purifier

Zhonglin Xu

Air Purifier

Property, Assessment and Applications

Springer

Zhonglin Xu
China Academy of Building Research
Beijing, China

Translated by Bin Zhou

ISBN 978-981-13-2531-1　　ISBN 978-981-13-2532-8 (eBook)
https://doi.org/10.1007/978-981-13-2532-8

Library of Congress Control Number: 2018954599

Translation from the Chinese language edition: 空气净化器—特性、评价与应用 by Xu Zhonglin,

This Springer imprint is published by the registered company Springer Nature Singapore Pte Ltd.
The registered company address is: 152 Beach Road, #21-01/04 Gateway East, Singapore 189721, Singapore

Preface

Theory, technique, measurement and equipment on pollution control of indoor particles (including microorganism) has been an important research area of Center for Cleaning Air Conditioning Technology from Institute of Building Environment and Energy at China Academy of Building Research (the former Air Cleaning Studio from Institute of HAVC at China Academy of Building Research), where the author worked and has been responsible for more than fifty years. The research in this field has been strengthened especially after the epidemic of "SARS" (Severe Acute Respiratory Syndrome). Many concepts have been proposed, which includes the corresponding theory and measurement of dynamic isolation (which is different from the past static isolation), the concept for preventing the entrance of dust particles and microbes into the room or the system (which is equivalent to the ideology for preventing the enemy from entering into the gate of the country), the concept of preventing and removing bacteria instead of killing bacteria, the combination of sealing and inducing, and the concept of sealing with dynamic air flow. A large amount of work related to the control for the process of microbial pollution has been performed by author, our research group in the air cleaning studio, and related organization and hospitals in Beijing. A series of patents (including international patents and national patents including these granted from USA) and products have been developed, including magnetic-type ultra-low resistance and high-medium efficiency air filter, effective air exhaust device sealed with dynamic air current, leakage-free exhaust air grille, outdoor air cleaning unit with self-cleaning function, energy-saving and anti bacterial air conditioning, and safe exhaust apparatus with online leakage scanning. Related standard in Beijing has been manipulated.

Recently, after the occurrence of smog, a systematic study on the pollution control of indoor air with air purifier has been initiated. This is based on the theoretical thinking for pollution control of particles (including microbes) with the research and practice of the electrostatic purifier in the late of 1960s, the furniture-type air purifier and the ultra-violet disinfector by shielded recirculation air in 1980s, and the air cleaning curtain in 1990s and the portable air purifier (abbreviated as air purifier hereafter). This is also started with the feeling for the incompleteness of foreign related standard. The content of this book is the result of the pre-research work of the sub-project No. 4 “Key-point technique and equipment for whole-process control of indoor microbial contamination” by special project entitled with “Green Building and Construction Industrialization” (No. 2017YFC0702800) by 2017 National Science & Technology Key Research and Development Program during the 13th Five-year Plan Period.

In recent years, the application of the air purifier has been developed rapidly. However, the delimitation on the performance of related products such as the so-called “Clean Air Delivery Rate” may cause divergence during theoretical analysis and evaluation of the product. People may be bewildered by the chaos from the product promotion, which result in the confusion for consumers during the selection of air purifier.

With the application of the fundamental principle of air cleaning technology, several problems related to air purifier are explained and clarified. It is hoped to be helpful to the correct design and application of air purifier.

Improvement to a certain extent can be realized with air purifier to reduce the harmful effect from indoor air pollution. But there is a “degree” for any problem. Then how to define this degree? It is better to analyze in theory and then validate it through actual case. This book is finalized through systematic theoretical analysis, calculation, experimental analysis and actual comparison. As the result of the pre-research work by National Science & Technology Key Research and Development Program, this book is aimed to attract readers’ interest, which is also expected to be useful for the study on indoor pollution control and the development of the air purifier industry.

Acknowledgement is also sent to Assoc. Prof. Dr. Bin Zhou at Department of HVAC in Nanjing Tech University, who is responsible for the translation of this monograph.

Beijing, China
August 2017

Zhonglin Xu
From Xu Zhonglin Studio
Institute of Building Environment and Energy
China Academy of Building Research

Contents

Chapter 1
Introduction

1.1 Development of Air Purifier

1.1.1 General Introduction

Air purifier was developed to reduce the airborne pollutants indoors which mainly include particles.

In China, the earliest air purifier developed was electrostatic cleaner. It was originally invented by the author's group from the air cleaning studio of the former Institute of Air Conditioning at China Academy of Building Research as auxiliary equipment in 1971 in Tianjin to reduce the particle concentration of the vortex area in cleanroom with mixed flow. It was termed as the electrostatic air purifier at that time.

Figure 1.1 shows the structure of the JZQ-I electrostatic cleaner developed based on the principle of single ionization region abroad. Figure 1.2 illustrates the structure of the JZQ-II electrostatic cleaner developed by author with the concept of "ionization with double regions" [1]. Under the condition of the same size, the turbidity efficiency of the former case (including the foreign prototype) reached 70–80%, while that of the latter case can reach 99%. In 1978, the national science and technology award was granted.

Household or public air cleaner developed for the first time in China was patented by author in 1988 with the title "*furniture with air purification system*". It is the air cleaner with mechanical barrier type. Although the performance of trial in some places was quite better, this product has not been promoted for application because of the insufficient understanding of environmental pollution. Later air purifying screen, i.e. the screen type air cleaner, was developed in Tianjin Purification Equipment Co., Ltd. with the entrustment of the editorial board of "Clean Technology".

It is well-known that UV irradiation can be used to kill bacteria, but occupants are not allowed to be exposed to it sometimes. For example, the ward for patients in recovery period after bone marrow transplantation, the laboratory for treatment

Z. Xu, *Air Purifier*, https://doi.org/10.1007/978-981-13-2532-8_1

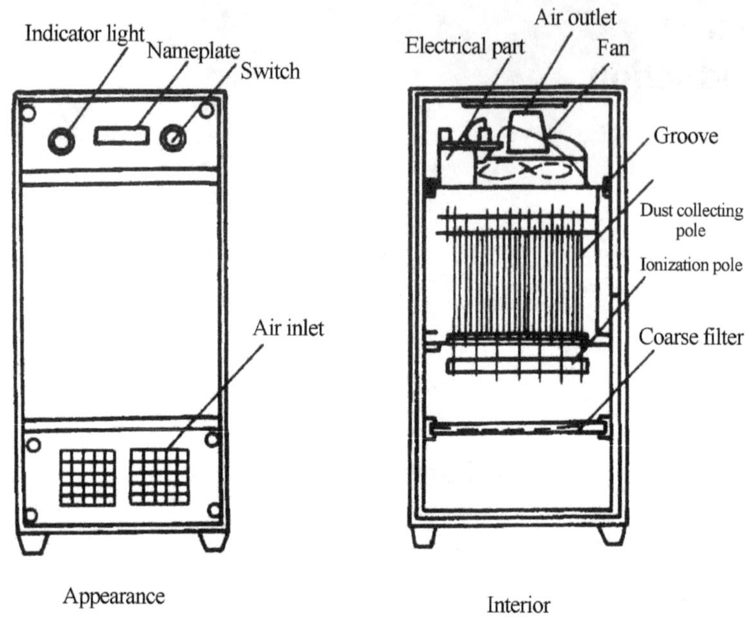

Fig. 1.1 JZQ-I electrostatic cleaner

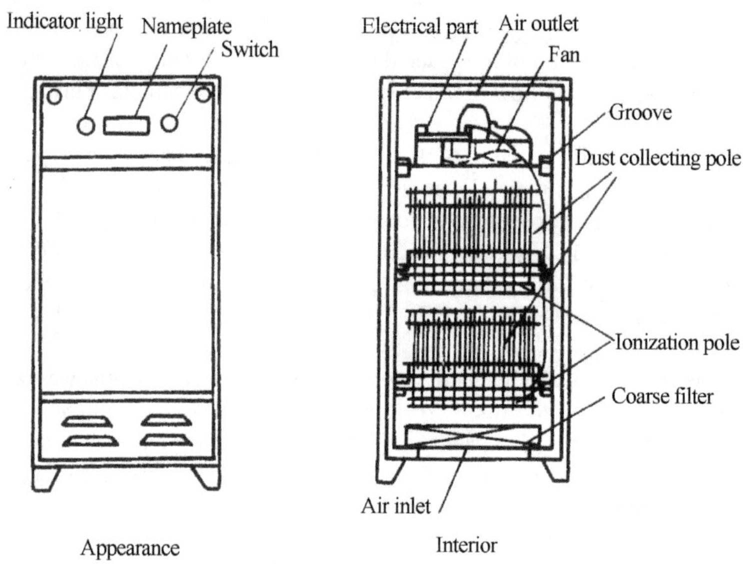

Fig. 1.2 JZQ-II electrostatic cleaner

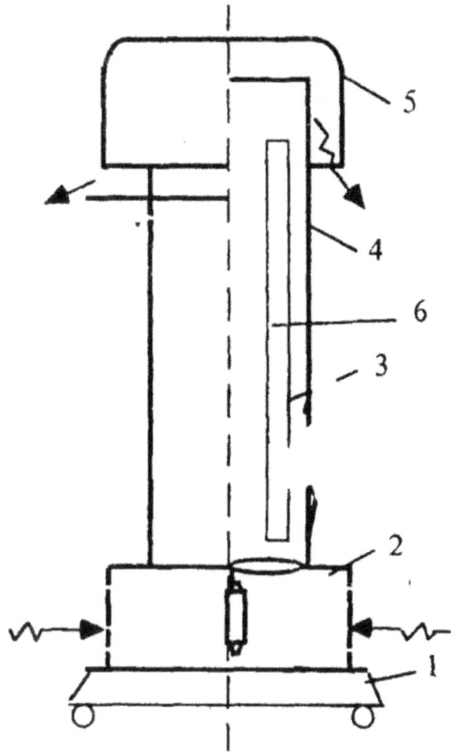

Fig. 1.3 Ground-type cylindrical UV sterilizer with circulating air. 1—movable base; 2—inlet air box; 3—disinfection region; 4—wind proofing ring; 5—light proofing cap; 6—UV lamp

of bone marrow, and the operating room for collection of bone marrow should be disinfected. At that time, the air cleaning technology has not been promoted in hospitals. Therefore, patients and staffs should leave during the disinfection process indoors. After several hours of disinfection with UV lamp, stop irradiation and start ventilation for a certain period of time. Then occupants are allowed to enter. It is inconvenient and the performance is rather poor. Besides, there are also other shortcomings for UV irradiation, which will be explained in detail later.

In order to make disinfection with UV lamp available during ordinary work period, the patented product for sterilization in the air stream was developed firstly in 1992 in China, which was entitled with *Shielded UV sterilizer with circulating air* (inventors include Chang-yong Chen, Zhong-lin Xu, Bing-yue Lin and Li-da Xu) [2]. There are two kinds of appearance design including ground-type (XK-1 type) and hanging type (XK-2 type), which are shown in Figs. 1.3 and 1.4, respectively.

It was pointed out by the novelty search report for the R&D product issued by Institute of Scientific and Technical Information of China on Feb. 16th of 1992 that among forty seven million literatures collected, there was neither literatures nor patents which contain the same content as the current project "*Shielded UV sterilizer with circulating air*" [2].

Fig. 1.4 Hanging type UV sterilizer with circulating air. 1—inlet air region; 2—disinfection region; 3—fan

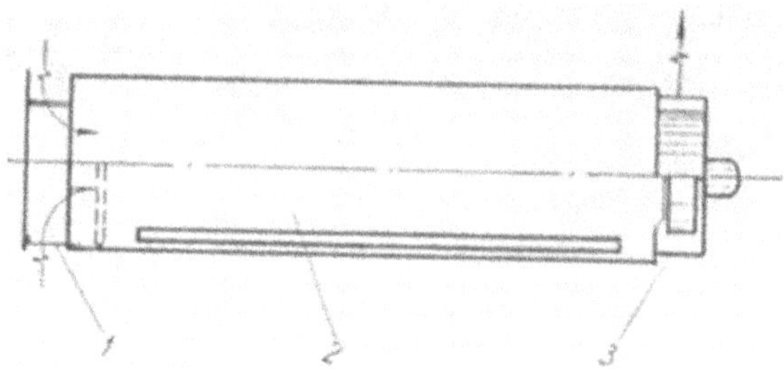

This product is one of the air purifiers for disinfection with UV irradiation. The key point during the design process of structure is the leakage prevention of UV irradiation with shielding measures.

The height of this air purifier is about 1.3 m, and the inner diameter is 264 mm. The flow rate is 354 m^3/h. The air velocity inside the cylinder is 0.4 m/s.

In order to prevent the spillover of UV irradiation, the shielding enclosure is adopted. Fan is installed inside to circulate air. In order to prevent the rapid loading of particles on the surface of UV lamp, as well as to reduce the particle concentration for large particles in the indoor air, coarse air filter is set at the entrance.

Since the application range of disinfection with UV irradiation is expanded with this product, popularity is received from users. It has been included in the chapter "*Disinfection of indoor air inside hospitals*" of the "*Technical Standard for Disinfection*" compiled by Ministry of Health of the People's Republic of China in 2000.

1.1.2 Developing Stage

Afterwards, there are three stages for the development of air purifier in China:

The first stage was the same period of "SARS" (Severe Acute Respiratory Syndrome) in 2003. The above-mentioned shielded UV sterilizer with circulating air played its role on the spot of "SARS". Meanwhile, various kinds of hanging type UV sterilizers appeared in the market. The episode of "SARS" promoted the sale of foreign electrostatic cleaner in China, which also promoted the development of domestic electrostatic cleaner.

The second stage was the period during the Olympic Game in 2008. Air purifiers with various principles appeared, which were aimed to be used in the Olympic gymnasium.

The third stage was the period for the outbreak of haze three years ago, which has attracted intensive care from the society. Domestic brands appear in the market sequentially, while foreign brands also rushed into Chinese market. According to "*2015 Trends Report for Air Purifier Industry in China*" from Electronic Information Industry Website, the sale volume of air purifier in China has already exceeded 3.2 million units. The amount of retail sale approached 7 billions, which was increased by almost 80% [3].

Haze is indeed the promoter for the development of air purifier, especially for Beijing-Tianjin-Hebei and surrounding areas. Although the land area of Beijing, Tianjin, Hebei, Shanxi, Shandong and Henan accounts for only 7.2% of the country, the coal consumed in these regions accounts for 33% of all the country. The emission intensity per unit area is about four time of the average national level. The production output of the steel per year in the region reaches 340 million tons, accounting for 43% of the country. That of the coke reaches 210 million tons, accounting for 47% of the country. The output of the electrolytic aluminum per day accounts for 38% of the country. The production output of the plate glass per day in the region reaches 12 million tons, accounting for 33% of the country. The production output of the cement per day in the region reaches 12 million tons, accounting for 19% of the country. The motor vehicle population accounts for 28% of the country (all the data above come from the interview of the minister Ji-Ning Chen at Ministry of Environmental Protection of the People's Republic of China on Jan. 6th of 2017). Therefore, although in terms of the national level, the $PM_{2.5}$ concentration in these three years was reduced by more than 30%, and the condition in Beijing-Tianjin-Hebei region was also improved. However, the average and peak concentrations were still high. According to the mid-term assessment report in "*the Air Pollution Prevention and Control Action Plan*" issued on Jul. 5th of 2016 by Chinese Academy of Engineering, among 338 cities at prefecture level and above, the conditions complied for only 73 cities. The ratio of days with concentration exceeded on average reached 23.3%. The ratio of days with severe pollution and worse case was 3.2%, of which 67.4% occurred in the winter. The days with severe pollution in Beijing-Tianjin-Hebei region accounted for 44.1% of the country.

Moreover, according to "*2016 Bulletin on China Environmental Statements*" issued by Ministry of Environmental Protection of the People's Republic of China in Jun. of 2017, the ratio of the days with pollution exceeding standard values reduced to 21.2%, and the number of cities with pollution within the standard limit increased to 84. It should be noted here that for the case with severe pollution and worse case, the days with $PM_{2.5}$ and PM_{10} as the main pollutant occupies 83.4 and 15.3%, respectively (in 2016 these two ratio values decreased correspondingly). Therefore, during the application of air purifier, people become realized that it is the priority that air purifier should have better purification performance on particles.

The standard related to pollution is related to the design and application of air purifier. Table 1.1 shows the related standards from several countries and World Health Organization (WHO).

In order to facilitate the evaluation of $PM_{2.5}$ pollution, the industry standard HJ633-2012 "Technical Regulation on Ambient Air Quality Index (on trial)" was issued by Ministry of Environmental Protection of the People's Republic of China. The requirement for $PM_{2.5}$ is shown in Table 1.2.

The limit of the so-called "Excellent" condition is 35 $\mu g/m^3$, which is still larger by than 10 $\mu g/m^3$ from the standard value of air quality given by WHO. The so-called overproof means the value exceeding 75 $\mu g/m^3$. The so-called severe pollution means the value exceeding 250 $\mu g/m^3$. The commonly-known "extremely high air pollution readings" is the value larger than 500 $\mu g/m^3$.

Table 1.1 Air Quality Guideline (AQG) value in WHO and some countries: Annual averaged concentration/24 h concentration

	PM_{10} (μg/m³)	$PM_{2.5}$ (μg/m³)
WHO: Target during transient period-1(IT-1)	70/50	35/75
Target during transient period-2(IT-2)	50/100	25/50
Target during transient period-3(IT-3)	30/75	15/37.5
Air quality guideline (AQG) value	20/50	10/25
US (implemented on Dec. 17th, 2006)		15/35
Japan (issued on Sep. 9th, 2009)		15/35
EU (issued on Jan. 1st, 2010, implemented on Jan. 1st, 2015)		25/None

Table 1.2 Classification limit for $PM_{2.5}$ concentration

Air quality index (AQI)	Index level of air quality	Type of air quality index and indicating color		Average gravimetric concentration of $PM_{2.5}$ within 24 h (μg/m³)
0–50	1st class	Excellent	Green	≤35
51–100	2nd class	Good	Yellow	>35–75
151–150	3rd class	Slightly polluted	Orange	>75–115
151–200	4th class	Moderately polluted	Red	>115–150
201–300	5th class	Heavily polluted	Purple	>150–250
>300	6th class	Seriously polluted	Maroon	>250–500

1.2 Classification of Air Purifier

1.2.1 Based on Treatment Object

According to the treatment object, air purifiers can be classified as follows

Particulate air purifier
Gaseous air purifier
Mixed air purifier (including particulate and gaseous).

1.2.2 Based on Application Field

According to the application field, air purifiers can be classified as follows

Household air purifier
General air purifier
Personalized air purifier (or the table top air purifier).

1.2.3 Based on Installation Method

According to the installation method, air purifiers can be classified as follows

Portable (or movable including hanging-type) air purifier
In-duct air purifier.

1.2.4 Based on Principle

(1) Principle for application of particulate air purifier mainly includes:

Filtration—applicable to all the particles (non-bacterial and bacterial)
Adsorption with high voltage electrostatic method—applicable to all the particles, especially for particles with diameter less than 1 μm
UV irradiation—only applicable to bacterial particles
Nano photocatalysis—only applicable to bacterial particles
Plasma and negative ion—mainly applicable to bacterial particles, and also capable of accelerating the deposition of particles.

(2) Principle for application of gaseous air purifier mainly includes:

Physical adsorption—such as activated carbon. Sometimes it may be limited because it may emit hazardous material during regeneration process.
Nano photocatalysis—reaction occurs mainly on the surface.
Chemical catalysis and complexation—there must be chemical substances, such as the chemical complex agent for absorption of chemical substances.

1.2.5 Applicability

According to the above-mentioned "trend report", the available brands of air purifiers in the market mainly cover almost 400 fields. "Principles" are diverse, which are even more than the afore-mentioned classification types. It includes adsorption + electrostatic, adsorption + filtration, adsorption + electric field, adsorption + electric field +

filtration, filtration + adsorption + catalysis + negative ion, chemical complexation, chemical complexation + filtration, electrostatic + photocatalysis + adsorption, electrostatic + photocatalysis + filtration + negative ion, etc. Moreover, components with these principles are put in series in the structure. Sometimes the air purifier contains component with even 10 principles.

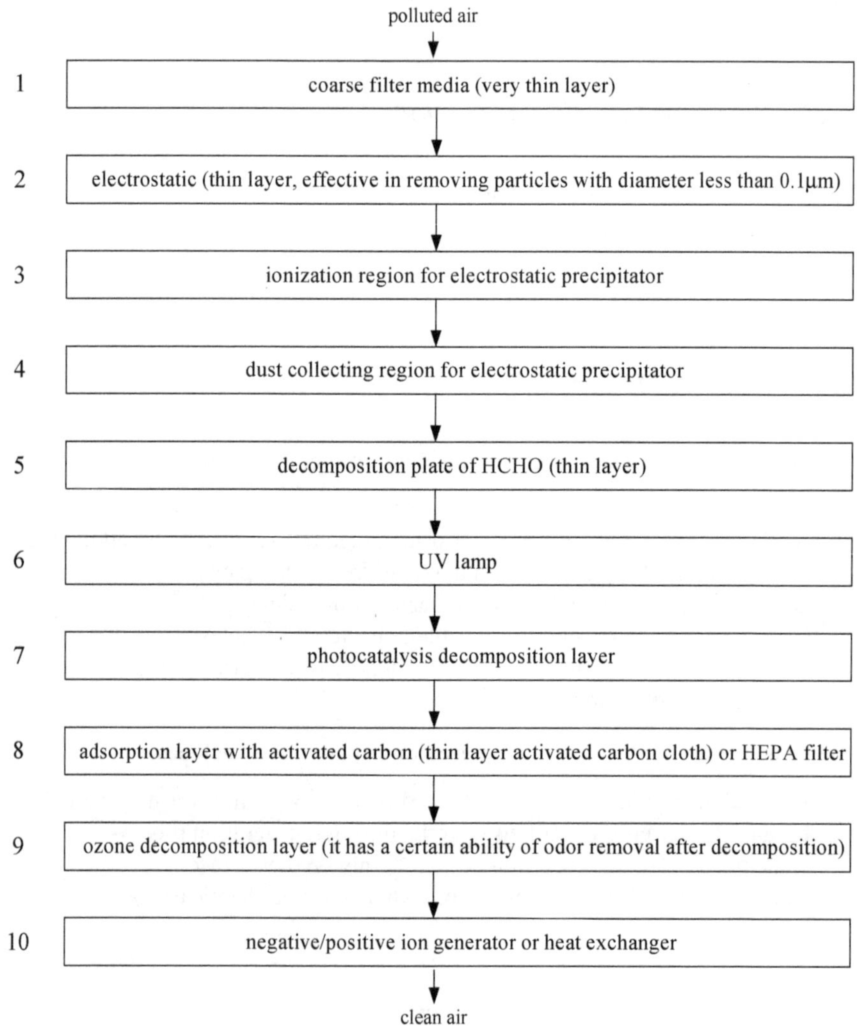

Some product samples are full of speculation in concepts. For example, the fashionable concept "nano" was speculated in nano-photocatalysis and nano-high frequency pulse light etc. Nano-titanium dioxide particles are used to form a coating. Under the irradiation of ultraviolet light with a specific wavelength, the so-called reactive oxygen species are generated inside the pores of the coating, which has a bactericidal effect. In fact, ozone may be generated because of ultraviolet light with different wavelengths. It is ozone which kills bacteria. But when dust is deposited onto the coating of the surface, the bactericidal effect disappeared. A big accident has been occurred before.

Table 1.3 lists various methods of air cleaning. The disinfection efficiency can be found in product sample, literature and test report from the manufacturer [4].

In the table, there is negative value for efficiency. This is because particles (including microbial particles) deposited on the surface will be blown out again.

In the promotion brochure of some products, it contains the misleading information including "without wearing parts" and "without consumables". When the electrostatic air purifier is concerned, it seems that it is no need to replace the filter media frequently. However, when more particles deposit onto the ionization wire, wreckage will be caused on electrodes. It is also difficult to clean the deposited particles on these dust collecting plates which are only several millimeter in between. When they are not cleaned for a long period, it becomes more difficult to clean. In this case, it should be cleaned with the ultrasonic cleaning machine. In 1950s and 1960s, the electrostatic air purifier was very popular in Japan. This is because since 1965 there was HEPA filter available in the market in Japan. Before 1965, HEPA filter was bought from U.S. Therefore, the electrostatic air purifier was also applied in the operating room. Practical experience showed that the spray water jet from the nozzle with pressure 2 kg/cm^2 should be used to clean the electrostatic air purifier, which was specified in "*Air Cleaning Handbook*" of Japan in 1981 [5]. It was shown that the deposited dust is very solid. For household application, it is difficult to imagine how to clean the space between plates with only several millimeters. If it is not appropriately cleaned and the thickness of the deposited dust is very large, the efficiency will be reduced. The resistance will be increased and the corresponding flow rate will be reduced. In this case, it is more likely to discharge. This is different from the advertisement that it can be used forever.

For the above-mentioned air purifier with ten layers of component in series, how to clean and replace one layer in between? This is completely out of reality.

In this book, only the principle of particulate air purifier is discussed, including three commonly used methods, i.e., the electrostatic adsorption, UV irradiation and especially the barrier-type filtration.

U.S. Environmental Protection Agency (EPA) issued "*Guide to Air Cleaner in the Home*", which introduced the applicability and limitation of several air purifiers [6]. It is shown in Table 1.4.

It is shown from the table for the limitation of "particulate air filter" that it is much targeted to apply air purifier to remove small particles from polluted air, since haze belongs to sub-micron particles instead of "relatively large particles".

In 2015, ASHRAE published "*ASHRAE Position Document Filtration and Air-Cleaning Devices*". It was pointed out that "Data from refereed archival literature are used to form summary statements on performance as well as the positions with respect to specific technologies. One key statement is that, at present, there is only significant evidence of health benefits for porous media particle filtration systems." "A key position is that filtration and air-cleaning technologies are not recommended for use

Table 1.3 Disinfection efficiency of various methods

Disinfection method	Disinfection principle	Disinfection efficiency
Electrostatic with single ionization region	The high voltage electric field forms corona, which produces free electrons and ions. Bacterial particles are charged by collision and adsorption of electrons and ions, which are precipitated on the dust collecting plate by deposition. The performance for comparatively large particle and fiber is poor, which may cause discharge. The advantage is its capability to remove bacterial particle with small resistance. The disadvantage is the difficulty to clean and it is time consuming. Pre-filter must be installed. Ozone and nitrogen oxides may be produces, which forms secondary pollution	50% (Test efficiency reaches only 20% for some product)
Plasma	Under the effect of heating or strong electromagnetic field, the highly ionized electron cloud will be generated from the gas. The active radicals and rays have a strong broad spectrum of killing effect on microorganism. Dust particles cannot be removed	66.70%
Negative ion	Under the influence of electric field, ultraviolet, ray and water impact, negative ion can be generated by ionization of air. It can absorb dust particles and form heavy ions to be settled. The disadvantage is the secondary dust, so the performance is not very good in air conditioning system	68.20%
Herb fumigation	Traditional Chinese medicine	73.40%
Nano photocatalysis	Under the sunlight and UV irradiation, volatile organic compound vapor or bacteria will be decomposed and converted into CO_2 and water by oxidation on surface of the catalytic activated matter. Enough time is needed for the contact between the disinfected air and the catalytic matter. With the deposition of particles onto its surface, the performance will be reduced greatly. UV irradiation will also generate ozone. Even negative value of efficiency appears in experiment	75% (Test efficiency reaches only 30% or even negative efficiency appears for some product)
Formaldehyde fumigation	Chemical agent, which has been claimed carcinogenic	77.42%

(continued)

Table 1.3 (continued)

Disinfection method	Disinfection principle	Disinfection efficiency
UV irradiation	When it is applied in air conditioning system where the air velocity is very large, the dosage of irradiation on bacteria is very small, so the performance is poor. It can only sterilize for bacteria, but it is inefficient for removing particles. Ozone will be produced. It is claimed by WHO and EU GMP as the unacceptable method and it cannot be used as the final sterilization measures	82.90%
Electron sterilization lamp	Physical method	85%
Electrostatic with double ionization regions	The ionization electrode and the dust collecting plate are separated	90% (Test efficiency reaches only 60% for some product)
Ozone	Light blue gas with relative strong oxidation effect. The oxygen atoms produced by decomposition can oxidize, penetrate the bacterial cell wall and kill bacteria. It has a broad spectrum of kill bacteria, but it cannot be used to remove particles. There must be no occupants indoors during application. It may damage a variety of items. The performance on surface microorganism is very small. It is harmful for the respiratory tract. It is not advocated to use by report	91.82%
Ultra-low resistance and high-medium efficiency filter	Physical barrier method. The resistance is only about 10 Pa for the opening with common size, which is one third of coarse filter's resistance. But the efficiency is comparable to that of high-medium efficiency filter (70–80% for particles with diameter $\geq 0.5\ \mu m$). It is light weighted and easy for installation. There is no secondary pollution	92–98%
HEPA filter	Physical barrier with no side effect. Non-reusable. "*Technical Standard for Disinfection*" compiled by Ministry of Health of the People's Republic of China pointed out that only air filtration method can be used for sterilization of air in cleanroom. The resistance is large	99.9999–99.99999%

Table 1.4 A brief summary of air cleaning technologies adopted by EPA

Air cleaning technologies		Pollutants addressed	Limitations
Filtration	Air filters	Particles	Ineffective in removing larger particles because most settle from the air quickly and never reach filters
	Gas-phase filters	Gases	Used much less frequently in homes than particle air filters. The lifetime for removing pollutants may be short
Other air cleaners	UVGI	Biological	Bacterial and mold spores tend to be resistant to UV radiation and require more light or longer time of exposure, or both, to be killed
	PCO	Gases	Application for homes is limited because currently available catalysts are ineffective in destroying gaseous pollutants from indoor air
	Ozone generators	Particles, gases, biological	Sold as air cleaners, they are not always safe and effective in removing pollutants. By design, they produce ozone, a lung irritant

if they produce significant amounts of contaminants that are known or expected to be harmful for health." "Extreme caution is warranted when using devices that emit a significant amount of ozone as by-product of their operation, rather than as a method of air cleaning. These devices pose a potential risk to health." These opinions are consistent with that shown in Table 1.4 (all the information is cited from Ref. [6]).

References

1. Z. Xu, *Fundamentals of Air Cleaning Technology*, 4th edn. (Science Press, Beijing, 2014)
2. C. Chen, Z. Xu, B. Lin, Development report of UV-type sterilizer for shielded circulation air (material from appraisal meeting), 1993
3. Z. Chen, Z. Lin, Research on actual filtration efficiency of electronic air cleaners. Refrig. Air-conditioning **17**(1), 71–75 (2017)
4. Z. Xu, J. Shen, *Guidance on the Implementation of Technical Specifications for Hospital Clean Operating Department* (China Architecture & Building Press, Beijing, 2014)
5. Japan Air Cleaning Association. *Air Cleaning Handbook*. 1981
6. J. Shen, Y. Liu, J. Yan, Correct understanding of medical environment control technology. J. HV&AC **46**(6), 73–78 (2016)

Chapter 2
High-Voltage Electrostatic Air Purifier

2.1 Type of High-Voltage Electrostatic Air Purifier

One of the main differences between electrostatic air purifier and industrial electrostatic precipitator is the application of discharge by positive corona. The ionization electrode is positive. Secondly, the DC voltage exerted on the ionization electrode and the dust collector is very small. Tens of thousands of volts are unnecessary, instead only a few thousand volts to million volts are needed. The resultant ozone generated is much less than that from the industrial electrostatic precipitator.

2.1.1 Single-Area Ionization

Single-area ionization can be divided into two forms, including the single-zone and double-zone types.

For single-zone type, both the ionization electrode and the dust collector are on one plate (or inside one tube), which are shown in Figs. 2.1, 2.2 and 2.3 [1, 2]. Figure 2.1 shows the discharge with ionization wire. Figures 2.2 and 2.3 show the discharge with tip.

For the double-zone type of electric field, the ionization pole and the dust collecting pole belong to two separate parts. As shown in Fig. 2.3 [1], it shows the discharge with ionization wire. The situation for discharge with the board edge is shown in Fig. 2.4.

Z. Xu, *Air Purifier*, https://doi.org/10.1007/978-981-13-2532-8_2

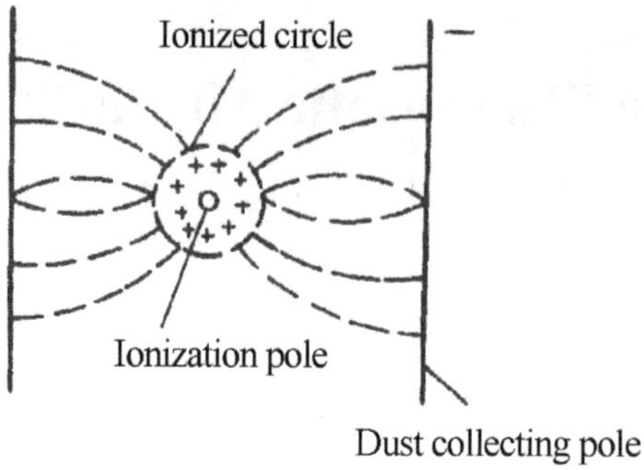

Fig. 2.1 Single-zone type of electric field

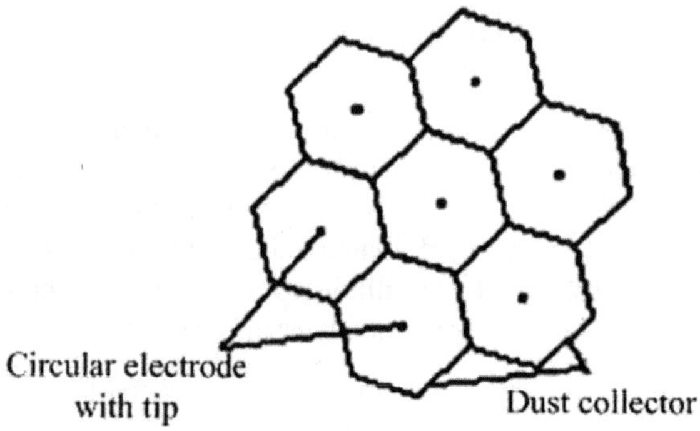

Fig. 2.2 Honeycomb tube electrostatic field (tip discharge)

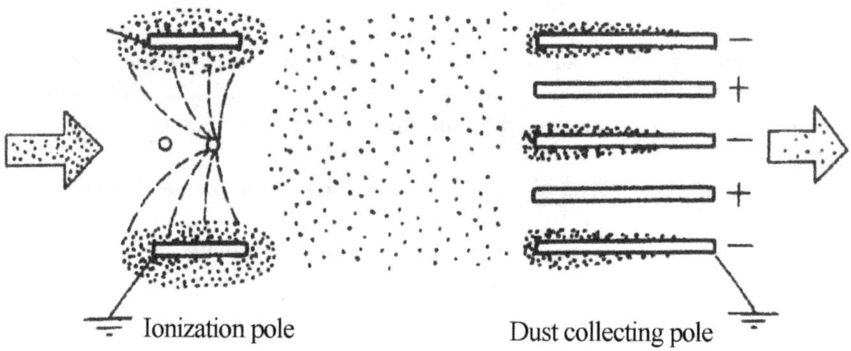

Fig. 2.3 Double-zone type of electric field

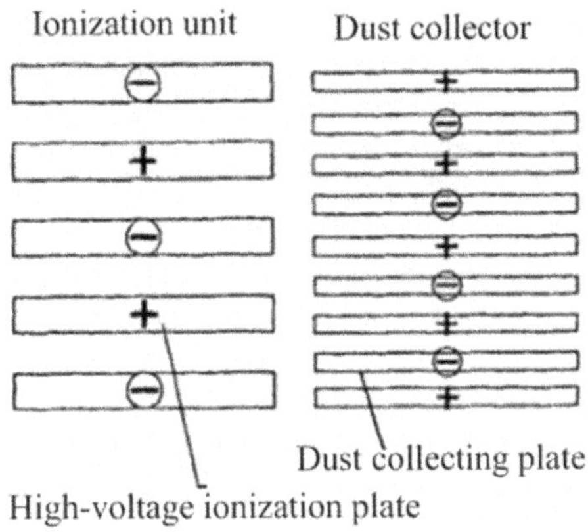

Fig. 2.4 Plate electrostatic field (board edge discharge)

2.1.2 *Double-Area Ionization*

As shown for JZQ-II type in Chap. 1, there are individual ionization electrode and dust collector in each area. There are two areas. Both ionization electrode and dust collector are placed in the first and second area, respectively. In this case, with the given length or height of the equipment, the dust collector is shortened (in terms of the dust collector, the proportion of particles collected on the rear half of the dust collector is very rare). Therefore, the sequence can be modified into:

$$\underbrace{\text{Ionization electrode} + \text{dust collector}}_{\text{1st area}} + \underbrace{\text{Ionization electrode} + \text{dust collector}}_{\text{2nd area}}$$

2.2 Main Principle and Structure of Electrostatic Purifier

2.2.1 *Main Principle*

When positive corona is used, high enough DC positive voltage is exerted on the metal wire of ionization electrode, and two sides of polar plates are grounded. In this way, non-uniform electric field is formed near the ionization electrode. A few free electrons in the air obtain energy from the electric field. They collide with air molecules fiercely, which generates collision ionization, and incomplete discharge appears, which is called corona discharge. Around the ionization electrode, a ring of light blue halo could be seen, which is termed as corona. So ionization electrode abounds with positive ions and electrons. Electrons move towards metal wire and neutralize on it. While positive ions move regularly under the effect of electric field, and they attached onto neutral particles when they encounter each other. In this way, particles become positive, which is the first kind of charge mechanism, i.e. electric field charge. Secondly, except for the movement under the effect of electric field, ions have thermal motion. Ions attach onto particles during the process of thermal movement, which makes particles positive. This is called the second kind of charge mechanism, i.e. diffusion charge.

After the positively-charged particles enter the dust collecting space (such as the space composed of plates with positive and negative electricity), they are repelled by the positive plate and then deposit onto the negative plate. The larger the voltage exerted on the dust collecting board is, the better the performance of repulsion and deposition is. Table 2.1 shows one examples of the result obtained by author.

In order to prevent the generation of ozone, electrostatic air purifier without high-voltage electrode is invented. Air entrance channel is made of special micro-porous material. Nano-carbon fibrous conductive layers are uniformly distributed inside. After electricity is connected, super electrostatic field is formed inside each micro-hole. It is used to capture particles, acting as the dust collector. But the efficiency of this kind of electrostatic purifier is very low. The initial efficiency is only 46% [3].

2.2.2 *Noticeable Problem in Structure*

It is very easy to cause the discharge with high voltage. During the discharge process, sparks and sound will be produced, which are dangerous. Meanwhile, ozone will be generated. The ability to capture particles will be reduced or lose. There was a case where the capture of particles was stopped in the field, and particles were resuspended and blown out downstream. The particle concentration of the supplied air increased rapidly, and some even caused combustion. Even if the surface of the plate has been treated with electro-polishing, the surface cannot be smooth for sure, especially near the edge where there is burr. Even though the surface is smooth, as long as there is one particle with large diameter on the surface, especially the fiber, the crackling sound can be caused. In some subway station, since the problem of setting Nylon mesh as pre-filter is not solved, people will not be dare enough to use the electrostatic purifier. Therefore, coarse or fine air filter must be placed at the entrance of inlet air before the electrostatic purifier.

Table 2.1 Relationship between efficiency and voltage of dust collector

Air velocity between dust collector plates (m/s)	Capacitor capacitance (μF)	Transformer output voltage (V)	Voltage of dust collector (V)	Turbidity efficiency (%)
1.3	8800	4200	7250	96.9
1.3	8800	3860	7000	96.5
1.3	8800	3410	6600	95.8
1.3	8800	2950	6250	93.7

In addition, the leakage through hold and gap inside the device should be prevented, which will reduce the efficiency. Two grounding plates should be added at the two edges of the ionization pole (electrode) and the dust collector (metal plate). Otherwise, the airflow and particles passing through the edges are not likely to be ionized and deposited, which reduces the efficiency.

Suppose particles can be effectively captured along the whole length of the dust collecting plate and the length is very long, the efficiency will approach 100%. But in reality not all the area of the plate along its length can collect dust efficiently. Test on JZQ-I electrostatic cleaner with length 30 cm shows that only 2/3 of the area along the length collect dust. If all the particles are deposited on the plate along this 2/3 length, the efficiency of this electrostatic cleaner approaches 100%, while in fact it is only 70–80% which can been seen from the comparison table about efficiencies. This is not caused by the length of plate which is not long enough so that particles do not have time to deposit, but by part of particles which is not charged or whose electric charge is not enough. For particles whose charge is not enough, the movement velocity u_e of the particle, i.e. the separation velocity, is small. It is helpful to prolong the length of dust collector plate for deposition of particles. However, for particles which are not charged at all, they will not be deposited on plate even when its length is prolonged. Since particles without charge do exist, a concept "*effective length of dust collector plate*" is proposed. It means that under certain electric field, only certain part of the plate has effect on dust collecting. When it is longer than this length, the longer part of the plate cannot be used to capture more particles or all the particles, as described by equation. The double-area ionization with shortened-length plate was proposed based on this principle.

Why do some particles carry very few charges or no charges? According to the theory of electric corona discharge, there are mainly two reasons:

(1) Since the electric ionization polar is a metal wire, the electric field with high electric strength only appears near the small distance around it, while the electric strength far from it is very weak. For the latter situation, the movement velocity of ions is very slow and the air in that region is not ionized (if all the air between plates is ionized, the electric field will be penetrated when spark discharge appears and short electric circuit is formed, thus the electrostatic cleaner stops).
(2) As mentioned before, with certain voltage of electric ionization polar, the ionization strength for air is fixed and the charge amount is determined. If the dust concentration of air entering electrostatic cleaner is high, charge on every particle is not enough or some particle can not be charged at all.

If the effective length of dust collector plate can be measured through experiment, the length of the plate can be shortened, or the double-area ionization scheme can be adopted.

The detailed explanation of principle can be referred to Ref. [1].

Table 2.2 Specification of product sample of air purifier during experiment

Sample No.	1#	2#	3#
Type of electric field	Double-area, wire and plate (discharge with electrode wire)	Single-area, channel with uniform circular hole (discharge with tip)	Single-area, plate (discharge with sawtooth type tip)
Dust collecting plate	Horizontal plate	Plate with circular hole	Small horizontal and parallel holes
Discharge voltage (kV)	14.0 (corona electrode) 7.0 (dust collecting plate)	7.0	7.0
Area of dust collecting plate (m^2)	5.376	1.414	1.250
Size (mm)	600 × 470 × 170	530 × 500 × 102	550 × 550 × 375
Effective area (m^2)	0.2820	0.2809	0.2860
Pre-filter	Metal wire pre-filter upstream	Metal wire pre-filter upstream and downstream	Non
Local photo of electric field in electrostatic purifier		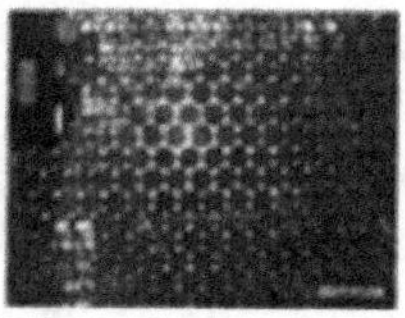	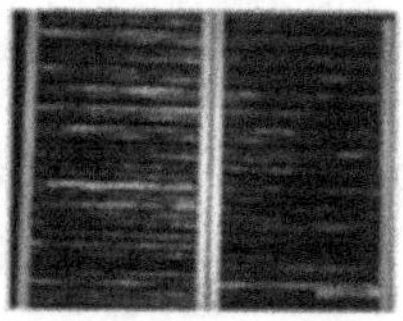

2.3 Efficiency of Electrostatic Purifier

2.3.1 Total Efficiency on Particles

(1) The theoretical calculation on efficiency of electrostatic purifier can be found in Ref. [1]. Measured data by Mao will be cited here [2]. Table 2.2 shows the specification of the product samples.
The efficiencies of the product samples in Table 2.2 are illustrated in Table 2.3. It is found that the efficiency of single-area type is too low. When the face velocity is between 0.8 and 1.8 m/s, the particle counting efficiency reaches 20–55%, which is only equivalent to that of coarse and fine filters (equivalent to G4-F6 filters based on foreign standard). The efficiency of double-area type can reach that of high-medium efficiency filter and sub-HEPA filter (equivalent to F7-F10 filters based on foreign standard). However, it was mentioned in Chap. 1 that the efficiency of double-area type can reach more than 99%.

(2) The voltage on the dust collecting plate is usually about 8 kV. Because a certain thickness of charged dust layer is formed on the dust collecting plate, the strength of the electrostatic field is weakened. After a period of usage, the efficiency reduces. Chen and Lin measured the performances of an imported

Table 2.3 Efficiencies of product samples in experiment

Fractional diameter (μm)		≥0.3	≥0.5	≥0.7	≥1.0	≥2.0	≥5.0
Flow rate (m^3/h)	Face velocity (m/s)	Particle counting efficiency with atmospheric particles (%)					
Sample No. 1							
1300	1.3	–	98.3	98.6	98.8	99.1	100.0
3400	3.3	–	76.8	87.8	93.5	97.0	100.0
Sample No. 2							
800	0.8	19.8	22.6	35.9	41.2	53.0	84.7
1200	1.2	14.1	16.1	23.7	29.3	50.0	69.4
180	1.8	7.6	9.0	15.8	21.0	37.5	58.2
2500	2.5	4.8	6.8	11.0	14.7	28.8	54.1
3000	3.0	3.9	5.7	9.9	21.5	36.8	38.2
3600	3.6	3.6	5.6	10.4	22.6	28.4	37.1
Sample No. 3							
800	0.8	38.1	41.4	48.8	54.8	68.6	75.0
1200	1.2	23.5	25.0	33.7	35.1	48.3	65.5
1800	1.8	16.2	17.7	21.9	26.8	35.0	43.3
2500	2.5	13.3	15.0	18.0	21.3	28.6	33.3
3000	3.0	9.0	10.0	11.8	13.9	16.5	28.9
3600	3.6	7.5	8.4	10.9	12.8	20.8	30.0

electrostatic air purifier with initial efficiency (≥0.3 μm) approaching 100% and an electrostatic purifier with non-high-voltage ionization polar with initial efficiency 46%. After continuous operation of one week (24 h/7 day), the efficiencies reduced. For the former sample, it reduced to 40% in the third week, and it was stable to about 35% after 15 weeks (including that for $PM_{2.5}$). Results are shown in Figs. 2.5 and 2.6 [4]. However, the efficiency of the latter sample only reached the half of the former sample. In the experiment, the thickness of the deposited dust layer was not provided. The performance after cleaning the dust layer was not measured. According to this experiment, if the device operates 10 h per day, cleaning process will be needed after usage of only five weeks. In order to prolong the period between cleaning processes, device to improve and stabilize the voltage of the dust collecting plate was included in some product. However, the structure will be complex in this case.

(3) With the increase of the air velocity passing through the dust collecting plates, the efficiency reduces. The measured result by author is given in Table 2.4.

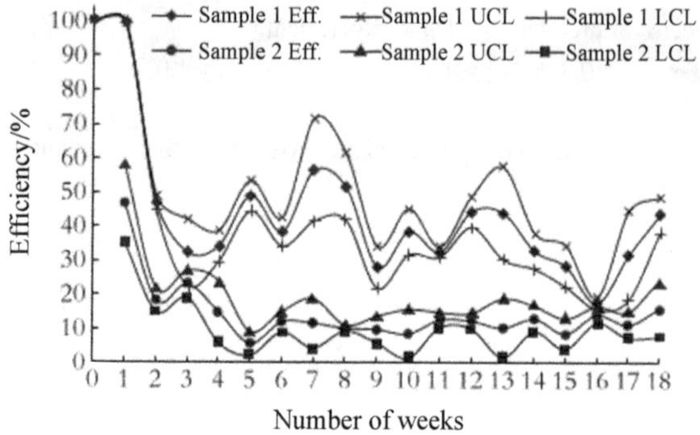

Fig. 2.5 One-pass efficiency with particle ≥0.3 μm for sample (Sample No. 1 is the imported electrostatic purifier; Sample No. 2 is the electrostatic purifier with non-high-voltage ionization polar. The initial efficiency of sample No. 1 is 99.9%. UCL and LCL mean upper and lower confidence limits, respectively)

Fig. 2.6 Efficiency with $PM_{2.5}$ for sample No. 1 (The initial efficiency is 99.9%)

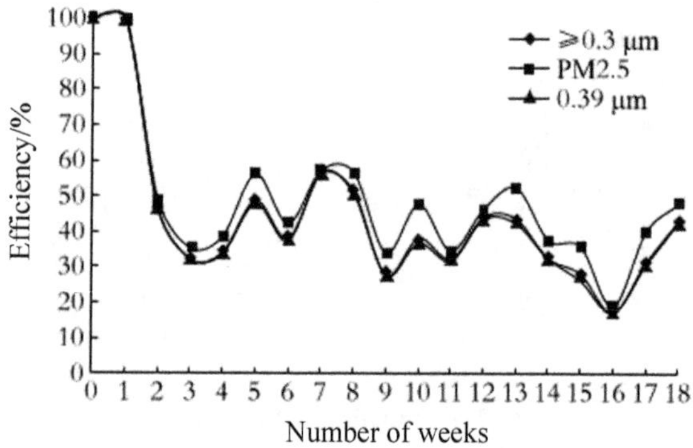

Figure 2.7 shows the performance curve of one product. The air velocity was added by author with the derivation of its size. When the face velocity became larger than 1.4 m/s, the efficiency reduced to be less than 90%. When the velocity was 2 m/s, the efficiency decreased to 70%.

Table 2.4 Influence of air velocity on efficiency

Air velocity through dust collecting plates (m/s)	Turbidity efficiency (%)
0.66	99.3
1.2	99.1
1.4	96.9
2.0	91.4

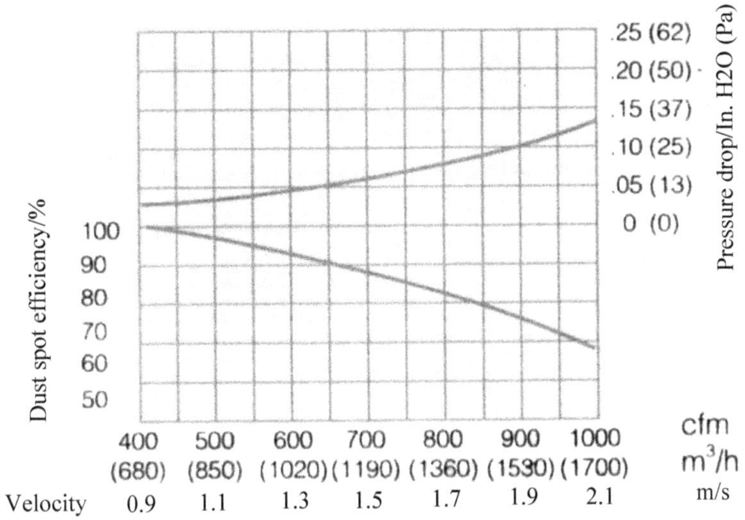

Fig. 2.7 Relationship between efficiency and velocity for one electrostatic air purifier (cited from the sample brochure)

2.3.2 Example of Bacterial Removal Efficiency

It is usually mentioned in the promotion brochure of electrostatic air purifier that bacteria can be killed. The mechanism of sterilization is usually related to high-voltage electric field. In fact, this is not the important property of electrostatic air purifier. This is because both bacteria and virus cannot survive alone (explained in detail later), and they live with the carrier. Since the electrostatic force eliminates the particles, this is equivalent to the removing of bacteria (since the original concentration of non-bacterial particle is largely different from that of bacterial particle, efficiencies on them are quite different). Without water or nutritious source, bacteria on the plate will die gradually as long as the electricity is not disconnected and they are not blown out.

Table 2.5 shows some measurement data for reference.

2.4 Attentions

(1) Safety

Except for the difficulty of cleaning mentioned before, electrostatic air purifier is the device with a nearly ten thousands of voltage. Because of the discharge caused by large particles and fiber, the safety problem cannot be ignored. After all, accident has been occurred in this aspect.

Table 2.5 Bacterial removal efficiency of electrostatic air purifier

Electrostatic air purifier (the type was not given)	After operation for 1 h	Bacterial removal efficiency with atmospheric bacteria reached 79.9% (45 m^3 room, 25–26 °C, 39–43%)	Master thesis from Tongji University, Mao [2]
	After operation for 2 h	Bacterial removal efficiency with atmospheric bacteria reached 91.1% (45 m^3 room, 25–26 °C, 39–43%)	
Electrostatic air purifier (single-area type)	After operation for 0.5 h	Bacterial removal efficiency in operating room reached 61.20%	Report from Institute of Microbiology and Epidemiology at the Academy of Military Medical Sciences, Yang [3]
		Bacterial removal efficiency in ICU reached 55.20%	
	After operation for 1.0 h	Bacterial removal efficiency in operating room reached 50.40%	
		Bacterial removal efficiency in ICU reached 73.20%	
	After operation for 1.5 h	Bacterial removal efficiency in operating room reached 57.00%	
		Bacterial removal efficiency in ICU reached 65.10%	
	After operation for 2 h	Bacterial removal efficiency in operating room reached 50.80%	
		Bacterial removal efficiency in ICU reached 86.50%	
	After operation for 3 h	Bacterial removal efficiency in operating room reached 78.80%	
		Bacterial removal efficiency in ICU reached 78.90%	
	After operation for 4 h	Bacterial removal efficiency in operating room reached 66.70%	
		Bacterial removal efficiency in ICU reached 78.60%	

(2) Ozone

The problem of ozone generation cannot be ignored. In order to avoid the generation of ozone at the expense of high voltage, the efficiency will be reduced.

(3) Dust Resuspension

Particles may be blown out during the power off and broken wire period. The device will be stop operating. The severe consequence can be imagined. Therefore, the mandatory provision was specified in National Standard GB50333-2002 "*Architectural technical code for hospital clean operating department*", which required that "electrostatic air purifier cannot be installed as the final air cleaning device of the clean air conditioning system". Later in 2008, the German standard on hospital DIN 1946 also specified that "electrostatic dust collector is not allowed to replace HEAP filter". This should also be noticed during the application of electrostatic air purifier.

(4) Efficiency Decrease

The rapid decrease of efficiency is a very severe consequence. This offsets the advantage of low resistance completely. The requirement on efficiency must be paid special attention to during application.

References

1. Z. Xu, *Fundamentals of Air Cleaning Technology*, 4th edn. (Science Press, Beijing, 2014)
2. H. Mao, Study on improvement of indoor air quality with electrostatic air cleaner, Tongji University, 2008
3. M. Yang, Significance of electrostatic sterilization in air cleaning, in *the 37th Pharmaceutical Preparations Forum & the 4th Pharmaceutical Disinfection and Sterilization Symposium*, 2009, pp. 127–128
4. Z. Chen, Z. Lin, Research on actual filtration efficiency of electronic air cleaners. Refrig. Air-conditioning **17**(1), 71–75 (2017)

Chapter 3
UV Irradiation

3.1 Feature of UV Irradiation

3.1.1 Advantage

At the beginning of the last century, the world's first water disinfection device with UV irradiation came into being. Later, it is well-known that UV irradiation with the C-band (wavelength 253.7 nm) was used to obtain the best performance of sterilization. Before the appearance of HEPA filter in 1950s, UV irradiation was the main means of gas sterilization. It has played an important role in purification and sterilization of indoor air. The general principle of UV light will not be repeated in this chapter.

The advantage of UV irradiation will be listed below [1, 2]. It is aimed to discuss the gain and loss during the application in air purifier.

(1) Broad spectrum of sterilization
Bacteria, fungi, virus, and even animal and plant cells can be killed under the exposure of the UV irradiation.
(2) Rapid sterilization
When the UV irradiation lamp is used within a short distance, rapid sterilization can be achieved within 1 s.
(4) It is especially applicable for emergency situation, such as the movable UV lamp.

Z. Xu, *Air Purifier*, https://doi.org/10.1007/978-981-13-2532-8_3

3.1.2 Disadvantage [1, 2]

(1) With sterilization capability only but enable to remove particles

Although bacteria are killed, the corps still exists. So the allergens still exist as non-bacterial particles.

(2) Selectivity of bacterial removal efficiency

Although UV irradiation has the broad spectrum of bacterial removal performance, the efficiencies for different kinds of bacteria differ a lot. The sensitivity of microorganism on UV rays can be found in Table 3.1 [3].

One of the reasons for different sensitivities of above-mentioned microorganism on UV ray is that the dose of UV irradiation to destroy the DNA on different cells is different.

If the exposure dose is defined as the product of the exposure intensity and the irradiation time, when the dose required for *Escherichia coli* is 1, the needed dose is about 1–3 for Staphylococcus, *Mycobacterium tuberculosis* and so on; it is about 4–8 for hay bacteria and its spore and yeast and so on; it is about 2–50 for mould fungi. Compared with the negative coli such as *Bacillus coli* and *Escherichia coli*, the sterilization rate for the positive coccal gambogic Sarcina is only 1/5–1/6, and the sterilization rate for positive bacillus such as the hay bacteria is only 1/11–1/14.

(3) Reversibility of sterilization

The bacteria killed can be resurrected again within 2 min in a short period or 1 min in a long period. Firstly, this "resurrection" is caused by the light resurrection. The visible

Table 3.1 Sensitivity of various typical microorganisms on UV rays

Sensitivity of on UV rays	Microbiological group	Microorganism of this kind
The most sensitive	Phytobacteria	*Staphylococcus aureus*
		Streptococcus pyogenes
		Escherichia coli
		Pseudomonas aeruginosa
		Serratia marcescens
↓	Mycobacterium	*Mycobacterium tuberculosis*
		Mycobacterium bovis
		Mycobacterium leprae
↓	Spore bacteria	*Bacillus anthracis*
		Bacillus cereus
↓		*Bacillus subtilis*
	Fungal spores	*Aspergillus versicolor*
		Penicillium chrysogenum
The least sensitive		*Stachybotrys chartarum*

light can activate the rejuvenated enzyme of the cell, which makes the replication of DNA successfully. Secondly, it is derived from self-repair. For example, under the action of various enzymes, the damaged parts of DNA molecule are removed and re-synthesized again [1, 2, 4].

(4) Stimulation for genetic mutation of bacteria

Bacteria can be mutated by UV irradiation. The most serious situation is the resultant drug resistance.

According to Sun's finding [5], after exposure to UV irradiation, both the *Enterobacter cloacae* and the *Staphylococcus epidermidis* became resistant to drugs. The eighth generation of the *Enterobacter cloacae* was found to have antibiotic resistance to the following five common antibiotics:

Ampicillin
Amoxicillin/boric acid
Cefoxitin
Cefazolin
Imipenem

The 5th generation of *Staphylococcus aureus* was found to be resistant to the following three antibiotics.

Ciprofloxacin
Erythromycin
Sulfamethoxazole compound

It was doubted in Ref. [6] that the proportion of airborne microbes with antibiotic resistance in hospitals was as high as 72–76.9%, which was much higher than that in the community. It is still unsure whether it is related to the frequent application of UV sterilization in hospital environment.

(5) Large influence by irradiation distance and exposure time

Irradiation distance. In the distance within range of 500 mm from the light tube center, irradiation intensity is in inversely proportional to the distance. But for the distance larger than 500 mm, irradiation intensity is in inversely proportional to the distance squared [7]. Figure 3.1 shows one example for the relationship between the irradiation intensity and the distance for a 15 W ultraviolet lamp. We can see from the picture that, when the irradiation intensity with the distance 100 mm is about 1200 $\mu W/cm^2$; it reduced to less than 600 $\mu W/cm^2$ for the distance 200 mm; it drops to 260 $\mu W/cm^2$ for the distance 400 mm; the irradiation intensity is about 120 $\mu W/cm^2$ for the distance 500 mm; and it reduced to one fourth of the former value, i.e. 30 $\mu W/cm^2$ for the distance 1000 mm; it drops to one fourth of the value for distance 1000 mm, that is 8 $\mu W/cm^2$ for the distance 2000 mm.

For a 30 W UV lamp, the UV irradiation intensities for different distances from its surface are shown in Table 3.2 [8].

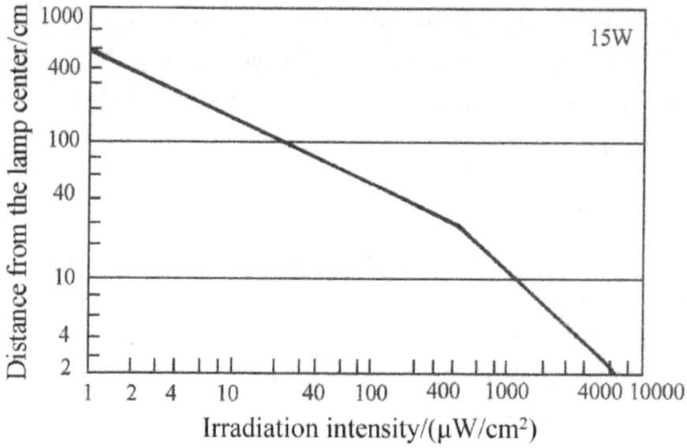

Fig. 3.1 Relationship between distance and irradiation intensity for a 15 W ultraviolet lamp

Table 3.2 UV irradiation intensities for different distances from lamp's surface

Intensity ($\mu W/cm^2$)	11,000	8900	7700	6400	4700	2900
Distance (mm)	0	10	20	34	65	100

It is found that when the distance is 10 cm from the lamp's surface, the intensity is reduced to 80% of the original value.

In order to obtain the irradiation dose needed, for the bacterial particle within a certain distance, it can only be achieved by prolonging the irradiation time. If the air velocity through the irradiation region by the UV ray is too fast, it can only be achieved by increasing the numbers of repeated passes. However, during this period, the effect of light resurrection cannot be ignored.

(6) Large influence by environmental factors

The output of UV lamp at 20 °C is the maximum, while it is reduced to 60% at 0 °C. The sterilization performance with the relative humidity 40–60% is the best. When the relative humidity is larger than these values, the sterilization efficiency decreases. If the relative humidity is more than 80%, the bacteria may even by activated.

(7) Reduced performance after continuous operation

The lifetime of UV lamp is generally more than 1000 h. The rated output of the UV lamp means the intensity after usage of 100 h. The initial output value is usually more than this value by 25%. The output after operation of 100–300 h will decrease gradually, which only reaches about 85% of the rated value.

(8) Generation of harmful gases

Harmful gases will be produced during the UV irradiation process. Nitrogen oxide and ozone will be generated for sure. The generated amount may be different for specific case. UV rays with wavelength less than 250 nm are more likely to produce photochemical smog and harmful gases.

Although the amount of ozone generated from the afore-mentioned cylindrical UV sterilizer is only 7 ppb which is far less than the background value, the nitrogen oxides are increased to more than 5 times of the background value after 3 h. This consequence by UV irradiation cannot be ignored.

3.2 Efficiency of UV Irradiation Purifier [1, 9]

3.2.1 Dose of UV Irradiation

With the certain dose of UV irradiation, the survival rate of bacteria is assumed to be:

$$S = 10^{-m} \tag{3.1}$$

Then the sterilization efficiency should be:

$$P = 1 - S = 1 - 10^{-m} \tag{3.2}$$

where m is the exponential index.

The irradiation dose needed E_0 for the specific type of bacteria when S value is fixed can be found in Table 3.3 [1]. When the sterilization efficiency required should be 90% for *Serratia marcescens*, the irradiation dose should be 1030 μW·s/cm². Results from different literatures on the irradiation dose needed differ a little, but sometime they differ a lot.

3.2.2 Calculation of Cylindrical UV Irradiation Purifier

For cylindrical UV irradiation purifier, the number n of 15 W UV lamp can be expressed as:

$$n = \frac{mE_0Q \times 10^{-2}}{1.1D} \tag{3.3}$$

Table 3.3 The UV irradiation dose needed in air when the sterilization rate is 90%

Type of bacteria	*Serratia marcescens*	*Escherichia coli*	*Gambogic sarcina*	*Bacillus subtilis* (spore)
Irradiation dose E_0 in air (μW·s/cm²)	1030	1000	4930	11,500

Table 3.4 Calculated number of UV lamps in cylindrical UV sterilization equipment

Cylinder diameter D (m)	Actual flowrate Q (m^3/min)	Experimental Bacillus E_0 (μW min/cm^2)	Actual tube number n (n × W)	Calculated value of m	Theoretical disinfection rate with actual tube number (%)	Theoretical disinfection rate with 99% of actual tube number (n × W)
0.264	5.33	17.1	3 × 30 (Consideration with 6 × 15)	1.91	98.8	7 × 15

where,

Q is the flow rate of the circulation air, m^3/min;
D is the cylinder diameter, m.

Based on Eq. (3.3), when the performance of three 30 W UV lamps is equivalent to that of six 15 W UV lamps, the value of m can be obtained. Then the value of P can be calculated with m. For the shielded UV sterilizer with circulating air mentioned in Chap. 1, the calculation results are shown in Table 3.4.

It is shown from the above table that for the UV purifier with three 30 W UV lamps, the final maximum theoretical sterilization efficiency can reach 98.8% when the flow rate is 320 m^3/h.

3.3 Measured Result of Bacterial Removal Performance

3.3.1 Distribution of UV Strength in Cylinder

When three UV lamps are set in this cylindrical UV irradiation purifier, the distribution of UV irradiation can be simulated by computer [8], which is shown in Fig. 3.2.

It is shown from the figure that the maximum intensity appears near the tube wall of the UV lamp, while the minimum value appears near the tube wall of the shell where it is equidistant from the two nearest lamps. There is a relatively flat high-dose area in the equilateral triangle with three UV lamps as the apex.

Since the length of the UV lamp is 1 m and the air velocity inside the cylinder is 0.4 m/s, the residual time is 2.5 s. Therefore, the irradiation dose in this high-dose area should be $2.5 \times 9000 = 22{,}500$ μW·s/cm^2. Here the effect of reflection by the bright tube wall is not included, which could reach 70%. Therefore, the sterilization efficiency should be higher.

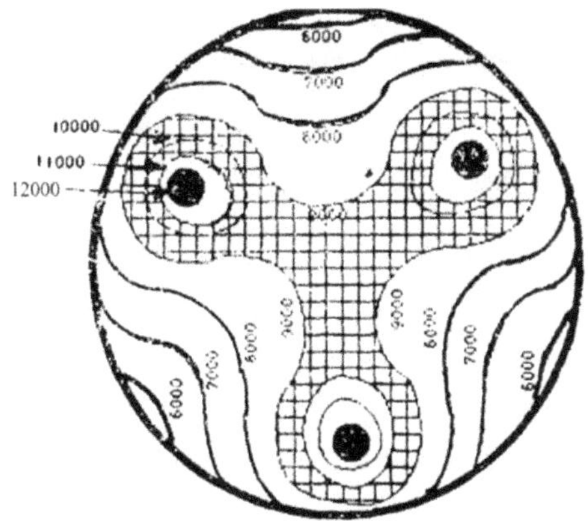

Unit of UV intensity/(μW/cm²)

Fig. 3.2 Distribution of UV irradiation on one cross-section of the cylinder

3.3.2 *Practical Measurement Result*

(1) When the afore-mentioned UV purifier was applied in a laboratory with the air change rate 11.2 h^{-1}, the sterilization performance is shown in Table 3.5 [8]:

It is shown that for the afore-mentioned XK-1 type cylindrical UV irradiation purifier, the sterilization efficiency did not reach the maximum value at 15 min. At 60 min, the purifier played it full role, when the maximum sterilization efficiency was obtained for the indoor air. This was also the theoretical sterilization efficiency with average value 97.8%. This is almost consistent with the calculated value 98.8% in Table 3.4. With the specular reflection from the cylinder, the theoretical efficiency should be higher, so the actual efficiency should also be larger. But with the protection of particles for bacteria, the actual performance will be less, which results in the equivalence of the measured and calculated data. After 60 min, due to the gradual decline of the overall number of indoor bacteria, the bacterial density through UV purifier will also be decreased. So the sterilization performance may be reduced.

(2) The afore-mentioned UV purifier was applied in the ward with area 10.84 m^2 (room height 2.6 m). Five sampling positions were placed on the working height. The disinfection performance is presented in Table 3.6[8].

Table 3.5 Sterilization performance in laboratory

Time (min)	Average bacterial removal efficiency measured for three kinds of room heights (%)
15	92.16–92.59
30	92.98–96.23
60	95.98–97.89
120	92.14–92.67

Table 3.6 Sterilization performance in ward

No.	Temperature (°C)	Relative humidity (%)	Number of people entering and leaving (Times)	Time (min)	Bacterial removal efficiency (%)
1	17.6	22	5	15	33.78–74.42
2	17.2	25	8	30	50.11–76.01
3	18.2	26	6	60	82.32–86.64

Table 3.7 Comparison of sterilization performance between different disinfection methods

Disinfection method	Bacterial concentration before disinfection (CFU/m^3)	Bacterial concentration after disinfection (CFU/m^3)	Bacterial removal efficiency (%)
Atractylodes rhizome	642	204	88.22
Ozone	673	55	92.82
Ultraviolet	601	103	82.86
Formaldehyde	598	135	77.42

Table 3.8 Sterilization performance of UV irradiation

UV lamp installed inside duct	Air velocity (v=3 m/s)	One-pass bacterial removal efficiency 80% [11]
UV sterilizer with circulating air was installed in a room with area 11.6 m^2	Outdoor air was supplied by unit with coarse, fine and sub-HEPA filtration systems	Indoor sterilization efficiency reached 93% after stable operation[a]

[a]Data was from the measurement result by the National Center for Quality Supervision and Test of Building Engineering

Data from Table 3.7 [10] and Table 3.8 are almost similar.

The application performance in the actual ward will be lower than that in laboratory. One reason is the entrance and exit of people. The other reason is that in laboratory measurement is performed with artificial *Serratia marcescens*. However, bacteria are attached to particles in ward, which may be protected by particles. It will also be influenced by non-uniform air distribution. Therefore, the measured result in ward with occupants inside is smaller than that in laboratory. The bacterial removal efficiency of UV irradiation, which can only kill bacteria but is enable to remove particles, is lower than the means which can remove both particles and bacteria.

References

1. Z. Xu, *Fundamentals of Air Cleaning Technology*, 4th edn. (Science Press, Beijing, 2014)
2. X. Yu, Characteristics of ultraviolet germicidal irradiation and application analysis. J. HV&AC **40**(7), 58–62 (2010)
3. M. Liu, J. Shen, C. Liu, Application of ultraviolet irradiation sterilization to ventilating and air conditioning systems. J. HV&AC **40**(4), 66–70 (2010)
4. S. Adachi, Microbicidal effects of ultraviolet radiation on environmental microorganisms. J. Antibact. Antifung. Agents **17**(1), 15–21 (1989)
5. R. Sun, M. Wang, H. Gao, X. Chang, The Effect of ultravoilet irradiation on *Aerobacter cloacae* and *Staphylococcus epidermidis*. Chin. J. Disinfect. **26**(2), 158–161 (2009)
6. Y. Chen, H. Li, Y. Lu, and L. Huang, Distribution of bacteriological spectrum in the air of hospital and analysis of their drug resistance. Chin. J. Antibiot. **31**(8), 505–506 + S2 (2006)
7. M. Huruhashi, Current study status of ultravoilet sterilization irradiation. Japan J. Med. Equip. **6**(7), 315–326 (1990)
8. L. Xu, B. Lin, J. Wang, Y. Wang, Design of shielded circulating air UV disinfection device and its performance. Chin. J. Public Heal. Eng. **2**(3), 132–135 (1993)
9. Z. Xu, C. Chen, J. Shen, Calculating method for the necessary lamps and sterile rate in a tube-shaped ultraviolet air washer. J. Hyg. Res. **27**(3), 213–216 (1998)
10. C. Wu, M. Du, Observation on performances of four air disinfection methods. Chin. J. Nosocomiol. **10**(6), 403 (2000)
11. Y. Wang, Study on dynamic disinfection techniques with in-duct ultravoilet germicidal irradiation, Tianjin University, 2011

Chapter 4
Fibrous Layer Filter

4.1 Principle of Fibrous Layer Filter

Granular layer, fibrous layer, microporous membrane, membrane covered fibrous layer, screen mesh and perforated plate can be used to remove particles, which are classified as barrier-type filter. This chapter only discusses the fundamental principle of fibrous air filter.

Fibrous layer comprises a fibrous filling layer, nonwoven fabric and filter paper etc.

There are mainly five principles for filtration of particles with fibrous air filter.

4.1.1 Interception (or Contact/Hook) Effect

Inside the fibrous layer, fibers locate randomly and form numerous meshes. The number of meshes formed could be dozens or even hundreds of layers. When particles with certain size approach fiber surface when following the streamline, suppose the distance between the streamline (also the centerline of particle) and fiber surface is equal to or smaller than particle radius, particles will be intercepted and then deposit on the fiber surface, which is called interception effect. Sieve effect also belongs to the interception effect.

4.1.2 Inertial Effect

Because of the complex arrangement of fibers, streamline will change directions frequently and abruptly when air flows through fibrous layers. When the particle mass is large or its velocity (which could be approximated by air velocity) is large,

Z. Xu, *Air Purifier*, https://doi.org/10.1007/978-981-13-2532-8_4

by virtue of its inertia, particle will not follow the streamline when air flows around fibers. Therefore, particles will deviate from the streamline when approaching fibers, then collide with fibers and deposit on it. If with the inertial effect, particles do not collide with the frontal surface of fiber but collide within the range of the interception effect, it is two combined effects which capture particles.

4.1.3 Diffusion Effect

It is much obvious for small particles that show Brownian motion caused by the collision between gas molecules and particles.

At normal temperature, the diffusion distance could be 17 μm/s for 0.1 μm particles, which is several times or even dozens of interspace distance between fibers. This could result in the larger possibility of particle deposition on fiber surface. While the Brownian movement is weaker for particles with diameter larger than 0.3 μm and usually it is not strong enough to make particle deviate from the streamline and deposit onto fiber surface.

4.1.4 Gravitational Effect

Particles will deviate from the streamline under the influence of gravity when they pass through fibrous layers, which mean they will deposit on fiber surface by gravitational deposition.

The time usage for airflows through fibrous air filters, especially paper filters, is far less than 1 s, so particles with diameter smaller than 0.5 μm will pass through fibrous layers without deposition on fiber surfaces. Therefore, it is reasonable to ignore the effect of gravitational deposition.

4.1.5 Electrostatic Effect

By various reasons, charge may exist on both fibers and particles, which will cause electrostatic effect and attract particles.

Except for the case of intentional charge on fibers or particles, both fiber charges by friction during the manufacture process and particle charge by charge induction will not exist for a long time. These electric field intensities are very week and the resultant attractive force is too small to be neglected.

With the combined effect of these five principles, there is the minimum efficiency for different kinds of particle diameter or filtration velocities. This is because sometimes the smaller the particle diameter is, the larger the efficiency is, while it is opposite for other principles. The effect of filtration velocity is also similar.

The quantitative analysis of these five principles and theoretical calculation on efficiency can be found in "*Fundamentals of Air Cleaning Technology*" [1] in detail.

4.2 Basic Index for Fibrous Layer Filter

There are four fundamental indexes of the fibrous layer filter.

4.2.1 Face Velocity and Filtration Velocity

Face velocity is defined as the air flow velocity passing the cross section of air filter (m/s). It represents the capacity of the flow rate passing through and installed area of air filter.

Filtration velocity is defined as the flow velocity of the air passing through the expanded area of filter media. It represents the ability of air flow passing through filter media, especially the filtration performance of filter media.

Table 4.1 shows the range of filtration velocity for air filters.

The magnitude of filtration velocity through coarse filter could reach m/s, but that of fine filter and high-medium efficiency air filter is usually dm/s. The range of filtration velocity for sub-HEPA filter is 5–7 cm/s, and that of HEPA filter and ULPA filter is usually 2–3 cm/s.

4.2.2 Efficiency and Penetration

When weight concentration is used to describe the particle concentration in the air flow, performance is evaluated with arrestance. When particle counting concentration is used, performance is evaluated with particle counting efficiency (such as microscopic counting and light scattering counting). When other physical parameter is used, performance is evaluated with dust spot efficiency or turbidity efficiency etc.

Table 4.1 Range of filtration velocity for air filters

Type	Coarse filter	Fine filter, high-medium efficiency air filter	Sub-HEPA filter	HEPA filter
Magnitude of filtration velocity	m/s	dm/s	cm/s	cm/s

The most commonly used method is the particle counting efficiency with particle concentrations at both the inlet and the outlet air flow, i.e.

$$\eta = \frac{N_1 - N_2}{N_1} = 1 - \frac{N_2}{N_1} \tag{4.1}$$

where N_1, N_2—particle concentration at inlet and outlet air flow (pc/L), respectively.

During the performance test on air filter, it is customary to use penetration K, instead of efficiency.

$$K = (1 - \eta) \times 100\% \tag{4.2}$$

According to author's calculation [1], under the condition of common distribution of atmospheric dust, when the efficiency for particles with diameter 0.3 μm is 99.91%, the corresponding efficiency for particles with diameter 0.5 μm is 99.994% and that for particles with diameter ≥0.5 μm 99.998%, which can be approximately considered as "Five 9" or 99.999%.

4.2.3 Pressure Drop

Pressure drop of air filter is composed of two components: filter media and structure of air filter.

The pressure drop of filter media is linearly proportional to the filtration velocity, while the whole pressure drop of air filter has exponential relationship with the filtration velocity, which can be seen in Fig. 4.1.

Except for reducing the pressure drop of filter media itself, there are other means to reduce the pressure drop of air filter:

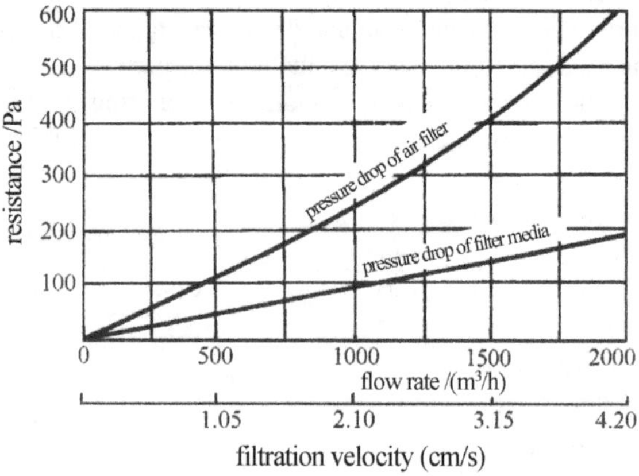

Fig. 4.1 Relationship between pressure drop and flow rate (filtration velocity) for HEPA filter

Expansion of area of filter media;
Reducing the length of air flow channel, this could reduce the air friction resistance;
Simplification of the entrance and exit of air flow channels, or reducing the air velocities at the inlet and the outlet, this could reduce the local resistance of air flow.

The above three measures are performed at the cost of increasing the filter area or volume. We can also try not to increase or try to increase a little of filter area or volume. There are already successful examples.

4.2.4 Dust Holding Capacity

Dust holding capacity is an index directly related to the life time of air filter. If air filter is allowed to use forever, particles will penetrate through the air filter and reach the leeward side of air filter, which is of course not permitted. Usually there is one restriction. In the early times, the dust weight deposited on air filter is called the dust holding capacity of this air filter, when the final pressure drop of air filter at operation is about two times of the initial pressure drop. It can be expressed as

$$\text{Final pressure drop} = 2 \times \text{Initial pressure drop at related air flow} \qquad (4.3)$$

Practice has shown that the larger the final pressure drop selected is, the larger the variation of the flow rate will be (the flow rate will be reduced by more than 50%). For the application where the flow rate should be kept constant, it is difficult to adjust the flow rate. Meanwhile, the energy consumption will be increased too much, which is not cost-effective. And pollution may also be caused (when the quality of air filter is not good, leakage or penetration is likely to occur by the increased pressure drop across the air filter. The gap inside the equipment may also increase the risk of leakage).

Both practice and theory prove that the operational flow rate through the air filter should not reach the rated value. By increasing the number of air filters, it seems that the operational flow rate of each air filter will be decreased. However, the life time of air filter will be prolonged more and the period between replacement cycles can be increased. The energy consumption by air filter during the longer operational period is reduced. With the reduced pressure drop, it is much energy-saving. Meanwhile, the efficiency is much better. Table 4.2 shows the relationship between the operational flow rate and the life time [1].

Therefore, it is specified in national standard GB50333-2013 "*Architectural technical code for hospital clean operating department*" that the operational flow rate of the final air filter should be larger than 70% of its nominal flow rate. The life time of air filter will be more than two times. For example, originally two air filters were used and they were replaced each half a year. In one year, they have been replaced twice, so six air filters have been used in total. Now when three air filters are used, they will be replaced once time during one year. Therefore, six air filters are also used in total. But since the pressure drop is smaller, it is much energy-saving. After operation of two years, there are ten air filters used for the former case, while only nine air filters are used for the latter case.

Therefore, air filter should not be replaced when the pressure drop increases to two times of the initial pressure drop at nominal flow rate. The flow rate will be decreased to a half of the initial value. It seems to be saving, but it is a waste of money in fact. Moreover, it is unacceptable when the flow rate is reduced to a half or even more. Now it is specified in the provision of some standard that air filter should be replaced when the final pressure drop reaches two times of the deigned initial pressure drop or two times of the operational initial pressure drop value.

That is

$$\text{Final pressure drop at replacement} = 2 \times \text{Designed initial pressure drop} \quad (4.4)$$

Or

$$\text{Final pressure drop at replacement} = 2 \times \text{Operational initial pressure drop} \quad (4.5)$$

They are given according to the above principle. Of course, the value of multiple can be adjusted according actual situation. For example, it is specified that air filter can be replaced when the efficiency becomes less than 85% of the initial value or the flow rate reduces to 85% of the initial value.

Moreover, since the filtration area in the air purifier should be as large as possible, its initial pressure drop is very small. According to Eqs. (4.4) and (4.5), the period to replace air filter may be short. A suitable multiple times of pressure drop can be provided by manufacturer by experiment. Then intelligent system can be used in air purifier.

For barrier-type fibrous air filter, the efficiency will not reduce but increase slightly during quite a long period (situation will be different when the deposited dust is too much). Therefore, it is reasonable to replace air filter when the flow rate is reduced to 85% of the nominal flow rate.

It is inappropriate to set the life time of the air purifier when the flow rate is reduced to a half of the nominal value. The standard value 85% commonly used in air cleaning industry should be adopted (for example, in GB50591-2010 "*Code for construction and acceptance of cleanroom*", it is specified that the difference between the operational and the designed flow rates of each air supply opening should not be larger than ±15% of the designed flow rate). This problem will be discussed again later.

Table 4.2 Relationship between the operational flow rate and the life time

Ratio between operational flow and nominal flow rate	0.5	0.7	0.75	0.8	1.0	1.25
Life time T_0 under nominal flow rate	$3.5T_0$	$2.15T_0$	$1.91T_0$	$1.7T_0$	T_0	$\frac{T_0}{1.7} = 0.59T_0$

4.3 Feature of Fibrous Layer Filter

4.3.1 Removing Both Dust and Bacteria

Fibrous air filter is effective in capturing particles.

Airborne particles can be classified as:

The total suspended particles (TSP): the airborne particles with aerodynamic diameter less than 100 μm;
The inhalable particles (IP), which is also termed as the floating dust: the airborne particles with aerodynamic diameter less than 10 μm; It can be expressed as PM_{10};
Fine particle: the airborne particles with diameter less than 2.5 μm; It can be expressed as $PM_{2.5}$;
Bacteria and virus itself are particles. The minimum size of naked virus is 8 nm, and the maximum of which reaches 0.3 μm. The size of naked bacteria is between submicrons to 10 μm. But both bacteria and virus must be attached to carriers. In this case, the size is called as the equivalent diameter.

Based on field measurement, when the density of bacteria is considered $\rho = 1$ g/cm^3 and $\rho = 1.5$ g/cm^3, the equivalent diameters for deposition of bacteria in ordinary and clean rooms are 5.2 and 3.9 μm, respectively. The equivalent diameter of virus is between 2 and 5 μm, which could be considered 3 μm [1]. For example, the real size of foot-and-mouth disease virus particle is only 25–30 nm. But cascade sampling results show that 65–71% of particles are larger than 5 μm, 19–24% are 3–6 μm, while only 10–11% are less than 3 μm [2].

Therefore, the opinion that fibrous air filter can only remove dust but cannot remove virus is not established.

It is shown from Fig. 4.2 that very few viruses can penetrate through air filter for most of viruses [3]. This means the efficiency is very high. Therefore, the efficiency for virus is not necessarily smaller than that for bacteria.

As for the worry about the penetration of microorganism through filter media or reproduction on filter media, people will consider to put some additives on the filter media for sterilization, which is unnecessary [1]. Taking the flu virus which people are worried about as an example, they can survive on surfaces of the stainless steel and the plastic for 24–48 h, but the survival period on surfaces of the paper and the handkerchief is less than 8–12 h [4]. This is because this kind of virus is not adapted to comparatively high relative humidity environment.

4.3.2 No Selectivity for Removing Dust and Bacteria

Dust particles include conductive and non-conductive, with water content or without water. Bacteria include plant bacteria, Bacillus, cocci, spore bacteria and fungi.

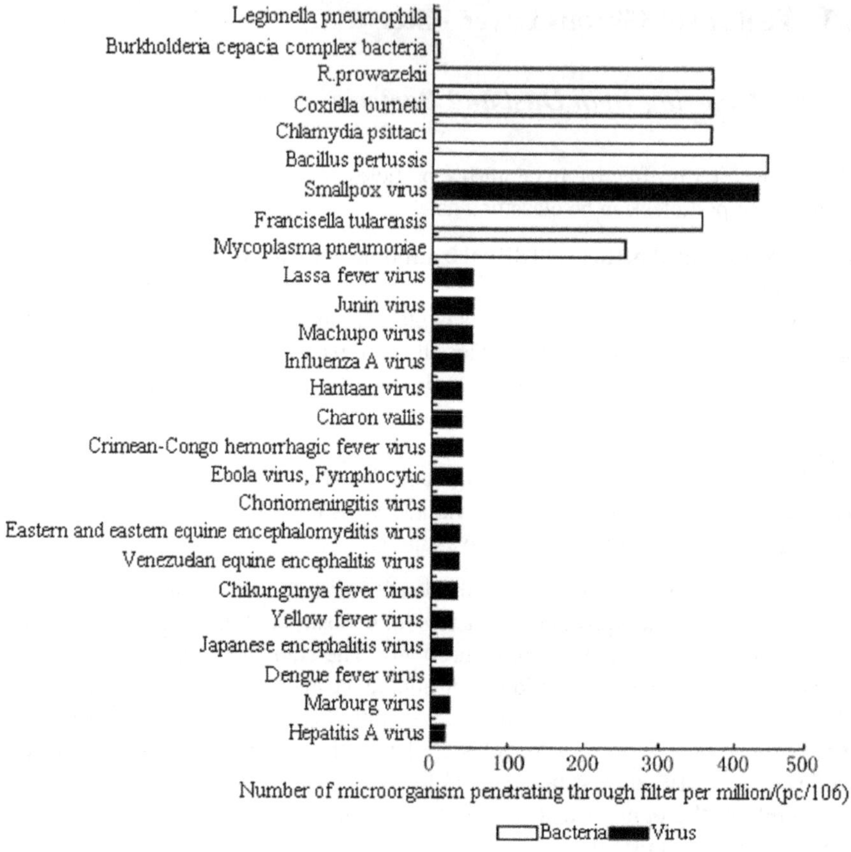

Fig. 4.2 Penetration of air filter for various kinds of microorganism through air filter

According to the ISO definition, fog is the general term for the droplet suspension in air. The combination of fog and a large amount of fine solid particles is called haze. In other words, haze contains water content (water, dissolved liquid of sulfate and nitrate etc.). High-voltage electrostatic adsorption is not a suitable means for removing conductive particles and water-containing particles.

Different bacteria have different sensitivity levels for UV ray, which can be seen in Table 3.1 of Chap. 3. Therefore, the sterilization performances of UV irradiation on different kinds of bacteria are different.

However, for barrier-type fibrous air filter, there is no selectivity for the "identity" of different particles. Instead, it only cares about the size of particles.

4.3.3 Broad Spectrum of Efficiency, Large Range and High Efficiency

Barrier-type fibrous air filters cover the range from coarse (the simple screen is not included) to ULPA filters, which is shown in Table 4.3.

The particle diameters used in test are different. For example, for HEPA filter or air filters with performance better, the Most Penetrating Particle Size (MPPS) should be used. For air filters with performance worse than HEPA filters, the particles with diameter 0.4 μm should be used. Therefore, there is a certain degree of relativity for the comparison table above. For the air cleaning equipment made in China, the name or symbol of air filter should be indicated according to Chinese national standard at first. Then the equivalent symbol with foreign standard should be indicated. It should not only indicate the latter symbol.

It is already shown in Chaps. 2 and 3 that the efficiency of air purifier with electrostatic and UV irradiation will usually decay gradually with the operational time. However, the efficiency of fibrous air filter will not decay but increase slightly, which could be considered constant.

It is already shown in Chaps. 2 and 3 that the efficiency of high-voltage adsorption and UV irradiation on particles or bacteria is within an extremely small range, such as "One 9" (90%). But the efficiency of HEPA filter on bacteria is larger than that for particles with diameter 0.3 μm or $\geq$0.5 μm. Table 4.4 shows the measurement result in negative pressure isolation ward [5]. The efficiency of HEPA filter on bacterial particles could be larger than that of particles by one magnitude.

This is because the linear length of the naked bacteria is about 1 μm and the width is about 0.5 μm, which is shown in Figs. 4.3 and 4.4 [6].

Spore has a very strong ability of resistance to ultraviolet irradiation, so the sterilization efficiency is relative low. But for air filters, it only cares about the size of the particles, regardless of whether there is spore or not. The bacterial removal efficiency of HEPA filter could reach "seven 9". It is incomparable by other methods.

Fig. 4.3 Electron microscopic photograph of *Bacillus subtilis* spores (Amplification magnitude 1700 times)

Table 4.3 Efficiency range of fibrous air filters

Chinese standard	Symbol	Efficiency E under nominal flow rate (%)		EU standard	
				EN779	EN1822
Coarse filter 4	C4	Standard artificial dust, arrestance	$50 > E \geq 10$	G1	
Coarse filter 3	C3		$E \geq 50$	G1–G2	
Coarse filter 2	C2	Particles with diameter ≥2.0 μm, particle counting efficiency	$50 > E \geq 20$	G2–G3	
Coarse filter 1	C1		$E \geq 50$	G3–G4	
Medium efficiency filter 3	Z3	Particles with diameter ≥0.5 μm, particle counting efficiency	$40 > E \geq 20$	G4–F5	
Medium efficiency filter 2	Z2		$60 > E \geq 40$	F5–F6	
Medium efficiency filter 1	Z1		$70 > E \geq 60$	F6–F7	
High efficiency filter	GZ		$95 > E \geq 70$	F7–F9	
Sub-HEPA filter	YG		$99.9 > E \geq 95$	F9	H10–H11
HEPA filter	A	Sodium flame efficiency	$99.99 > E \geq 99.9$		H12–H13
HEPA filter	B		$99.999 > E \geq 99.99$		H14
HEPA filter	C		$E \geq 99.999$		U15
ULPA filter	D	Particles with diameter ≥0.1–0.3 μm, particle counting efficiency	≥99.999		U16
ULPA filter	E		≥99.9999		U17
ULPA filter	F		≥99.99999		U17

Table 4.4 Efficiency of HEPA filter on particles and bacteria measured in negative pressure isolation ward

	Efficiency for particles with diameter ≥0.5 μm	Efficiency for *Bacillus subtilis* and spore
HEPA filter B	99.999	99.99997
HEPA filter C	99.99994	99.999997

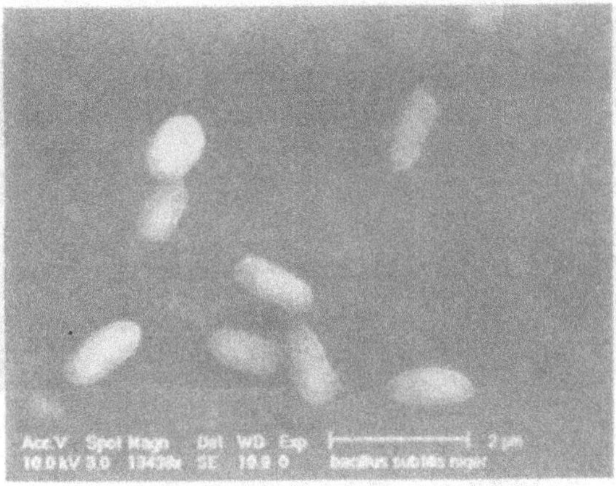

Fig. 4.4 Amplified electron microscopic photograph of *Bacillus subtilis* spores (Amplification magnitude 13,500 times)

Air filter such as HEPA filter is made of filter media paper with thickness 0.25–0.3 mm. The media is formed by 100 layers of fibrous network. Even though the particle capture efficiency for each network layer is less than 10%, the total efficiency may reach five 9. So it is not surprising for the high efficiency of HEPA filter.

4.3.4 No Side-Effect

The barrier-type air filter will not produce any harmful substance (particles or gases), electromagnetic field, or chemical reaction. It is a mechanical barrier.

In German standard DIN1946-4 (2008), it was pointed out for air cleaning device that "During the operation of the unit, it is not allowed for any harmful substances to volatile into the unit, even the use of antibacterial filter or filters containing special material". Air filters containing the antibiotics and the lysozyme belong to these special material. Therefore, during the episode of SARS, ASHARE Journal stated not to support the chemical filter. It repeatedly warned to carefully use antibacterial products. It has been pointed out in Chap. 2 that the afore-mentioned German standard does not allow to replace the HEPA filter with electrostatic filter, which has already specified in Chinese standard GB50333 in 2002.

Similarly, UV irradiation cannot be considered as the best assistant. Or even it is thought to replace air filter. The negative evaluation of ultraviolet (shown in

Table 4.5 Evaluation of sterilization method with UV rays by GMP at home and abroad

GMP	Provisions	Evaluation
WHO	17.34	"......Because of the limited performance of ultraviolet, it cannot be used to replace chemical disinfection method"
	17.65	"......UV irradiation must not be used as the final disinfection method"
EEC (European Economic Community)	Appendix 1:70	"......Usually UV irradiation cannot be used as the sterilization method"
EU (European Union)	Appendix 1:70	"......Usually UV irradiation is not an acceptable sterilization method"
China, 2010 version	Appendix 1:43	"......Ultraviolet disinfection cannot be used to replace chemical disinfection"

Table 4.5) by Good Manufacturing Practice (GMP) at home and abroad should attract our thinking [7].

So after literature review, Memarzadeh pointed out at last that ultraviolet germicidal irradiation (UVGI) is not recommended for air management before air recirculation from airborne isolation rooms. It also is not recommended as a substitute for HEPA filtration [4].

The above material and the evaluation by U.S. EPA cited in Chap. 1 provide the proof for the view stated in "*ASHRAE Position Document Filtration and Air-Cleaning Devices*" published by ASHRAE in 2015 [8]. It means that barrier-type fibrous air filter may be the only choice for air purifier. The above point of view is:

> This Position Document addresses the health consequences of filtration and air cleaning. Data from refereed archival literature are used to form summary statements on performance as well as the positions with respect to specific technologies. One key statement is that, at present, there is only significant evidence of health benefits for porous media particle filtration systems. For a few other technologies, there is evidence to suggest health benefits, but this evidence is not sufficient to formulate firm conclusions. A key position is that filtration and air-cleaning technologies are not recommended for use if they produce significant amounts of contaminants that are known or expected to be harmful for health.

4.3.5 *No Decreased but Increased Efficiency*

During the effective use of air filter, the efficiency usually does not decrease but increase with the increase of dust loading. Of course, when there are too much deposition of particles, the air filter is out of the period of effective use. This is different from the electrostatic purifier in Chap. 2.

4.3.6 Little Influence of Air Velocity on Efficiency of Equipment

Because the filtration area of the filter media can be far larger than the face velocity of the equipment, the influence of face velocity on efficiency could be reduced. The face velocity of electrostatic purifier is or slightly less than the air velocity passing through, so it is not applicable to the situation with larger amount of flow rate.

4.3.7 Relative Cheap Price

The unit with air filter or the air supply outlet with air filter is much cheaper and simpler than these with nano-photocatalysis, electrostatic adsorption etc.

4.3.8 Regular Replacement with Certain Pressure Drop Value

The fibrous air filter has the resistance, which is often considered as one factor in the criticism. As mentioned before, although the resistance of electrostatic purifier or UV irradiation or nano-catalysis itself is very small, they cannot be used alone. Coarse of fine air filter must be installed at the entrance, whose resistance at nominal flow rate could reach 50 Pa according to standard specification. In order to prevent the leakage of UV rays and particles, air filters with higher efficiency are usually placed at the exit of the UV germicidal irradiation unit. In one product, even HEPA filter is added. For safety reason, air filter or even HEAP filter is set at both the entrance and the exit. Table 4.6 shows the pressure drops of several electrostatic purifiers by Mao [9]. When W-type air filter was added downstream, the pressure drop was very high. Therefore, it is incorrect in the generally speaking that the pressure drop of electrostatic purifier or UVGI unit is less than fibrous air filter. The pressure drop of the whole unit should be used for comparison.

With the improvement of structure, the pressure drop of the fibrous air filter can be reduced greatly. Table 4.7 shows the pressure drop of ultra-low resistance and high-medium efficiency air filter invented by author, which has particle removing efficiency 85% ($\geq$0.5 μm) and bacterial removal efficiency more than 99%. It was measured by the National Center for Quality Supervision and Test of Building Engineering.

When the structures of air filter and equipment are improved, the structural resistance could be greatly reduced and the life time becomes longer. The new patented product is available now.

For the patented outdoor air cleaning unit in the II-type clean operating room, the coarse filter is self-cleaning and its resistance is extremely small, and there is no final pressure drop. The pressure drop in the section for fine and high-medium efficiency

Table 4.6 Measurement result of initial pressure drop for air purifier

No.	Flow rate (m^3/h) (Face velocity m/s)	Initial pressure drop of air filter (Pa)			
		Sample No. 1	Sample No. 2	Sample No. 3	Sample No. 3+W filter
1	800 (0.8 m/s)	2.9	2.9	2.9	170.6
2	1000 (1.0 m/s)	4.9	4.9	3.9	215.8
3	1200 (1.2 m/s)	7.8	6.9	4.9	274.6
4	1500 (1.5 m/s)	10.8	10.8	7.8	337.4
5	1800 (1.8 m/s)	14.7	13.7	11.8	419.7
6	2000 (2.0 m/s)	18.6	18.6	14.7	488.4
7	2500 (2.5 m/s)	27.5	29.4	22.6	635.5
8	3000 (3.0 m/s)	39.2	41.2	33.3	788.5
9	3600 (3.6 m/s)	55.9	59.8	49.0	994.4

Table 4.7 Pressure drop of ultra-low resistance and high-medium efficiency air filter

Air velocity (m/s)	0.31	0.41	0.52	0.61	0.71	0.79	0.89	0.99
Pressure drop (Pa)	8	10	12	13	15	17	19	21

Note Data were from the test report from the National Center for Quality Supervision and Test of Building Engineering and measurement result from Hong-hong Pan, Guo-qing Cao et al.

filter (or sub-HEPA filter) at nominal flow rate is only 14 Pa. Because the filtration area is expanded, the service life is greatly prolonged.

The pressure drop of sub-HEPA filter media is much less than that of HEPA filter media. With the appearance of HEPA filter media with novel material, the pressure drop could be reduced by one third.

As for the replacement of air filter, it is much convenient and safer than the replacement of components for the multiple layer and in-series air purifier introduced in Chap. 1, or the cleaning of electrostatic purifier. The cleaning of non-woven fibrous material will consume a lot of water and it is time-consuming. Moreover, the original performance will also be greatly reduced.

References

1. Z. Xu, *Fundamentals of Air Cleaning Technology*, 4th edn. (Science Press, Beijing, 2014)
2. X. Yu, F. Che, *Modern Air Microbiology and Sampling Detection and Identification Techniques* (Military Medical Science Press, Beijing, 1998)
3. X. Yu, Air cleaning: the major measure for removing microbial aerosol particles. J. HV&AC **41**(2), 32–37 (2011)

4. F. Memarzadeh, R.N. Olmsted, J.M. Bartley, A. Arbor, B. Hills, Applications of ultraviolet germicidal irradiation disinfection in health care facilities: Effective adjunct, but not stand-alone technology. Am. J. Infect. Control **38**(5 Suppl 1), S13–S24 (2010)
5. Y. Zhang et al., Experiment on germ filtering efficiency of high efficiency filters on return air inlet in isolation wards. J. HV&AC **36**(8), 95–96+112 (2006)
6. J.A. Schmidt, System and method of applying energetic ions for sterilization (2003)
7. H. Pan, Discussion on applicability of sustained using ultraviolet rays in air conditioned rooms for disinfecting and sterilization: Part 6 of the series of research practice of the revision task group of the Architectural technical code for hospital clean opera. J. HV&AC **43**(7), 27–29+36 (2013)
8. J. Shen, Y. Liu, J. Yan, Correct understanding of medical environment control technology. J. HV&AC **46**(6), 73–78 (2016)
9. H. Mao, Study on improvement of indoor air quality with electrostatic air cleaner, Tongji University, 2008

Chapter 5
Operational Property of Air Purifier

According to the analysis in aforementioned chapters, particulate air purifier is only needed for common environment. The priority of choice is fibrous air filter. In order to obtain the main basis for design of air purifier, the operational property of air purifier must be understood [1].

5.1 Operational Conditions of Air Purifier

According to the application purpose and situation, the operational conditions of air purifier can be classified as four categories.

5.1.1 In Test Chamber

In test chamber, there is no particle generation from the chamber cabin or inside chamber. The chamber is closed and there is no occupant inside. For the purpose of test, smoke is generated inside the chamber and then mixed to be uniformly distributed. The required particle concentration is obtained to be the initial concentration value. The self-circulation operation process of air purifier is then started. Once the required time period is reached, the air purifier will be turned off.

5.1.2 Conditions with Introduction of Partial Outdoor Air

There is particle generation indoors. In order to prevent the invasion of atmospheric dust and meet the hygiene requirement of occupant, a proportion of outdoor air is treated and then mixed with a large proportion of return air inside the air purifier,

Z. Xu, *Air Purifier*, https://doi.org/10.1007/978-981-13-2532-8_5

which is supplied indoors. In this case, most of the supplied air is the return air. This operational condition is common in room where there is no ventilation system.

5.1.3 Conditions with Complete Indoor Self-circulation

There is particle generation indoors. Meanwhile, there is invasion of atmospheric dust and no introduction of outdoor air. The air purifier is operated under the condition of self-circulation. This is the most common situation for operation of air purifier.

5.1.4 Conditions with Supplementary Function to Existing System

There is ventilation system indoors and even air cleaning system. Air purifier is placed locally to remove local vortex or increase the local air change rate, so that the purification effect at these local places will be improved. In some cases, the total air change rate of the whole room or system may be reduced. In short, it acts as a supplementary role for the existing system.

5.2 Operational Property in Test Chamber

5.2.1 General Formula for Steady-State Particle Concentration

Different allocation schemes of air filters in the ventilation system are shown in Fig. 5.1.

For these systems with different allocation schemes, expressions for indoor particle concentration are different, which are quite difficult to be remembered. In 1971, the consistent expression of the improved uniform distribution theory was proposed by author. After two novel concepts “Total efficiency on the pipeline of outdoor air” and “Total efficiency on the pipeline of return air” were adopted, the expression becomes unified and very concise. Therefore, the steady-state indoor particle concentration can be expressed as [2]:

$$N = \frac{60\mathrm{G} \times 10^{-3} + \mathrm{Mn}(1 - \mathrm{s})(1 - \eta_\mathrm{n})}{\mathrm{n}[1 - \mathrm{s}(1 - \eta_\mathrm{r})]} \tag{5.1}$$

where

η_n is the total efficiency of air filters on the pipeline of outdoor air;
η_r is the total efficiency of air filters on the pipeline of return air;

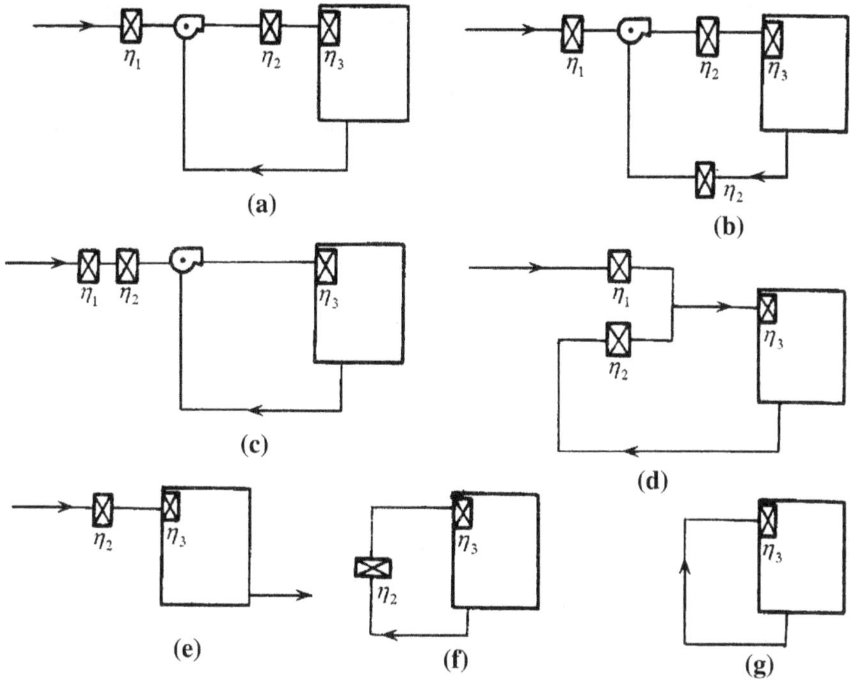

Fig. 5.1 Diagrams of systems with different allocation of air filters

G is the uniformly-distributed particle generation rate by indoor particle generation source, pc/(m^3 min); It can also be transformed into the particle mass concentration during calculation;
M is the concentration of atmospheric dust, pc/L; It can also be transformed into the particle mass concentration during calculation;
S is the ratio of return air;
n is the air change rate, h^{-1};

For case (a) in Fig. 5.1, we obtain:

$$N = \frac{60G\times 10^{-3} + Mn(1-s)(1-\eta_1)(1-\eta_2)(1-\eta_3)}{n[1-s(1-\eta_2)(1-\eta_3)]}\text{pc/L} \quad (5.2)$$

For case (c) in Fig. 5.1, we can obtain:

$$N = \frac{60G \times 10^{-3} + Mn(1-s)(1-\eta_1)(1-\eta_2)(1-\eta_3)}{n[1-s(1-\eta_3)]}\text{pc/L} \quad (5.3)$$

It is the same for the rest categories. If the unit of the particle concentration is converted into μg/m^3, the essence of the formula is not affected.

5.2.2 General Formula for Transient Particle Concentration

When $(1-\eta_1)(1-\eta_2)\cdots$ in the numerator is expressed as $(1-\eta_n)$ and those in the denominator expressed as $(1-\eta_r)$, the particle concentration N_t at time t becomes:

$$N_t = \frac{60G \times 10^{-3} + Mn(1-s)(1-\eta_n)}{n[1-s(1-\eta_r)]} \cdot \left\{1 - \left[1 - \frac{N_0 n[1-s(1-\eta_r)]}{60G \times 10^{-3} + Mn(1-s)(1-\eta_n)}\right] e^{-nt[1-s(1-\eta_r)]/60}\right\} \tag{5.4}$$

where

N_0 is the initial concentration indoors at time $t = 0$.

When Eq. (5.4) is expanded, we obtain:

$$N_t = \frac{60G \times 10^{-3} + Mn(1-s)(1-\eta_n)}{n[1-s(1-\eta_r)]} - \frac{60G \times 10^{-3} + Mn(1-s)(1-\eta_n)}{n[1-s(1-\eta_r)]} e^{-nt[1-s(1-\eta_r)]/60} + N_0 e^{-nt[1-s(1-\eta_r)]/60} \tag{5.5}$$

5.2.3 Expression of Particle Concentration Inside Test Chamber

Take case (g) in Fig. 5.1 as an example, it is similar to the case with the wall mounted air purifier, where air purifier is installed at the tuyere position and clean air is supplied. In this case, there is neither outdoor air pipeline nor circulated pipeline shown in the figure. Air is returned by the opening below the air purifier. If there is no particle generation indoors and the initial concentration is N_0, it is the condition in the test chamber. Equation (5.5) can be converted into Eq. (5.6).

$$N_t = 0 - 0 + N_0 e^{-nt[1-s(1-\eta_r)]/60} \tag{5.6}$$

This means:

$$\frac{N_t}{N_0} = e^{-nt[1-s(1-\eta_r)]/60} \tag{5.7}$$

Since it is self-circulation, $s = 1$. When the unit of n is $\min^{-1}$, Eq. (5.7) can be rewritten as:

$$\frac{N_t}{N_0} = e^{-n\eta_r t} \tag{5.8}$$

For this case, it was firstly proposed by ANSI/AHAM AC-1-2006 "*Method for Measuring Performance of Portable Household Electric Room Air Cleaners*" [3] to test the air purifier. It is specified that in the closed test chamber with given volume, there is no particle generation from the test chamber. After the mixture with the fan, the initial concentration N_0 (with smoke generator, $\geq 0.3\ \mu m$) is obtained. After the air purifier is turned on for time period t (such as 20 min specified in national standard related to air purifier), the indoor particle concentration reached N_t. Based on the measured data, the "decay constant" k in this period can be obtained. We can get Eq. (5.9).

$$\frac{N_t}{N_0} = e^{-kt} \tag{5.9}$$

The distribution by this equation conforms to the exponential variation trend of indoor particle concentration under uniformly-distributed ideal condition, which is shown in Fig. 5.2.

When the process of self-purification reached the point b and there is a sudden generation of indoor particle, the pollutant curve will be elevated from b to c′. When the particle generation stops at c′, indoor air will be further purified towards b′, which has the same concentration as the initial place b.

The value of k is the difference between the total decay process and the natural decay process within the specified test period. It is not discussed here.

When Eqs. (5.8) and (5.9) are compared, we obtain:

$$k = n \cdot \eta \tag{5.10}$$

where

η is the total efficiency of air filters inside the air purifier, which is η_r in previous sections.

However, there are also other factors influencing the value of k during measurement. Of course, the one-pass efficiency of air filters is still the main factor. The following description is based on this fact.

Fig. 5.2 Variation of particle concentration

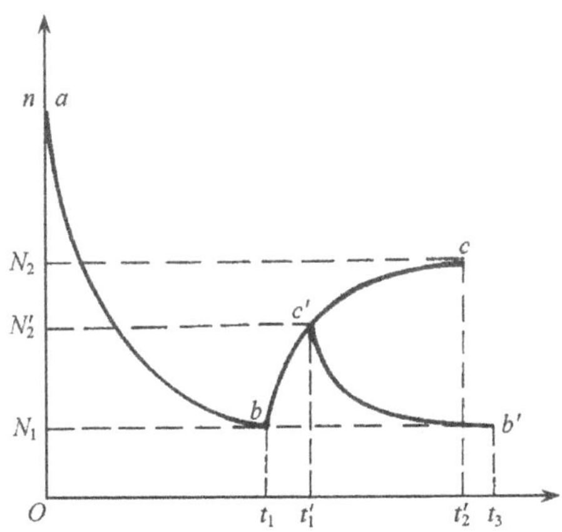

5.2.4 Discussion on Setting of CADR

The test method on air purifier was firstly specified by The Association of Home Appliance Manufacturers (AHAM) in U.S.A. There was no theoretical derivation. Instead, it was only an artificial definition with the so-called Clean Air Delivery Rate (CADR) Q to describe the "purification performance for the target pollutant", which was used to reflect the operational property of air purifier.

Q is defined as:

$$Q = 60\,\mathrm{kV} \tag{5.11}$$

where

V is the volume of the test chamber.

The capability of reducing indoor pollutant by air purifier depends on the flow rate which corresponds with the air change rate based on the room volume, as well as the efficiency of the components including air filter. Besides, the place of the air purifier, the shape and allocation of supply and return air openings will influence the air distribution, which will affect the purification performance. This can be reflected by the non-uniform coefficient in cleanroom. For air purifier in the test chamber, the value of k cannot be used to reflect the influence. This will not be discussed temporarily.

With Eqs. (5.10) and (5.11), Eq. (5.12) can be obtained..

$$Q = 60\,\mathrm{n}\eta\mathrm{V} \tag{5.12}$$

Since the nominal flow rate is $Q_0 = 60\,nV$, Eq. (5.13) can be derived based on Eq. (5.11).

$$Q = \eta Q_0 \tag{5.13}$$

In this equation, Q is set as a specified physical parameter—the flow rate. Moreover, it is termed as the parameter of flow rate for "clean air". It is used to show the purification ability. If it is increased by one times (η or Q_0 is increased), will the purification ability to reduce indoor particle concentration also be increased by one times? The positive answer is likely to be the consensus. However, the real answers for the aforementioned three conditions are quite different.

For the first situation in Sect. 5.1, if there are two air purifiers with the same efficiency η and $Q_{0b} = 2Q_{0a}$, we know $Q_b = 2Q_a$ based on Eq. (5.13). Does this mean the ratio of purification capability between b and a is 2:1?

Of course, the purification capability means the ability to reduce indoor particle concentration. The ratio of purification capability means the ratio of the reduced indoor particle concentration. Because there is no interior particle generation source inside the test chamber, the steady-state particle concentration will reach 0 when

$t \rightarrow \infty$. Therefore, the steady-state particle concentration cannot be used. Instead, the particle concentration N_t at time t can be used. This will be explained with the following example.

Equation (5.14) can be obtained from Eq. (5.8)

$$\frac{N_{at}}{N_{bt}} = \frac{e^{n_b \eta_b t}}{e^{n_a \eta_a t}} \tag{5.14}$$

Suppose $\eta_b = \eta_a = 0.9$ and $Q_{0b} = 2Q_{0a}$, we get $Q_b = 2Q_a$. From Eq. (5.12) and Eq. (5.13), we know $n_b = 2n_a$. Suppose $n_b = 0.2\ \text{min}^{-1}$ and $n_a = 0.1\ \text{min}^{-1}$, the ratio of particle concentration inside the test chamber in 20 min becomes:

$$\frac{N_{at}}{N_{bt}} = \frac{e^{n_b \eta_b t}}{e^{n_a \eta_a t}} = \frac{e^{0.2 \times 0.9 \times 20}}{e^{0.1 \times 0.9 \times 20}} = \frac{36.5982}{6.0496} = 6$$

This means the ratio of purification capability between air purifier b and a is 6:1, which is different from the ratio of the CADR between Q_b and Q_a. Although $Q_b{:}Q_a = 2$, the ratio of the purification capability is not 2:1.

Moreover, if there are two air purifiers with the same flow rate Q_0 and $\eta_b = 2\eta_a$, does this mean the ratio of the purification capability between b and a is 2:1 or 6:1?

This will also be explained with the following example.

Suppose the same flow rates Q_0 for two air purifiers are the same, the air change rates n are the same. Suppose $n_b = n_a = 0.1\ \text{min}^{-1}$. If $\eta_b = 0.999$, we get $\eta_a = 0.5$.

The ratio of particle concentration inside the test chamber in 20 min becomes:

$$\frac{N_{at}}{N_{bt}} = \frac{e^{n_b \eta_b t}}{e^{n_a \eta_a t}} = \frac{e^{0.1 \times 0.999 \times 20}}{e^{0.1 \times 0.5 \times 20}} = \frac{7.374}{2.718} = 2.713$$

This means the ratio of purification capability between air purifier b and a is 2.713:1, instead of $\frac{0.999Q_0}{0.5Q_0} \approx 2:1$ or 6:1.

From the above analysis, the expression of CADR cannot be used to reflect the first situation, where the variation of the purification capability is neither proportional to the CADR, nor proportional to n or η. It will cause confusion during assessment.

5.3 Operational Property in Conditions with Introduction of Partial Outdoor Air

5.3.1 General Formula for Steady-State Particle Concentration

When partial outdoor air is introduced, the second situation applies. Since there is only one air filter, the steady-state indoor particle concentration can be obtained by the non-uniform distribution theory [4]:

$$N = \psi \frac{60G \times 10^{-3} + Mn(1 - s)(1 - \eta)}{n[1 - s(1 - \eta)]} \quad (5.15)$$

where

G is the particle generation rate indoors, pc/(m^3 min);
n is the air change rate, h^{-1};
N is the particle concentration, pc/L;
η is the total efficiency of air filters in air purifier;
M is the concentration of atmospheric dust, pc/L;
S is the ratio of the return air to the total flow rate in air purifier;
Ψ is the non-uniformity coefficient. Based on the exiting research result on cleanroom [4], when $n = 2$–5 h^{-1}, $\Psi \approx 4$–2. There is no specific value for air purifier. But it is sure that the non-uniformity performance is worse than the airflow with the upper-supply and down-return scheme.

There is no real impact on the expression when the particle counting concentration or the particle mass concentration is adopted (particle counting concentration is used in the test chamber). It will be explained in detail in Chap. 8 as to how to choose suitable unit.

Although this situation is not specified in standard related to air purifier, it is also not specified that CADR is not applicable to this situation. In the market, there is such air purifier.

5.3.2 Relationship Between Efficiency, Air Change Rate and Particle Concentration

The concept of CADR means that the larger the value of CADR is, the greater the purification capability of air purifier is. It is easily misunderstood that CADR is proportional to the purification capability. With this concept, the shortcoming of lower efficiency can be counteracted by the increased flow rate. Take the air purifier with $\eta = 0.95$ as an example, in order to obtain the same CADR and the same purification capability as the air purifier with $\eta = 0.995$, Q_0 should be increased by $\frac{0.995}{0.95} = 1.05$ times. However, this result will not appear for the second situation.

For this situation, the flow rate of the outdoor air is not very large. The ratio of outdoor air is less than 0.3, so the minimum of s can reach 0.7. Suppose M is the concentration of the atmospheric dust during the haze episode, when it is seriously polluted, $M \geq 10^6$ pc/L (≥ 0.5 μm). It is shown in Table 5.1 [5].

Suppose $G = 5 \times 10^5$ (detailed calculation process can be found in Chap. 8). The total particle counting efficiencies ($\geq$0.5 μm) are 0.95, 0.995, 0.9995, 0.99995 and 0.999995, respectively. The air change rate n increases from 10 to 200 h^{-1}. When it is assumed to be uniformly distributed (Ψ is not considered at first), results are shown in Fig. 5.3 [6].

From the above figure, when $M = 10^6$, $n = 10\ \mathrm{h}^{-1}$ and $\eta = 0.95$, the CADR provided can reduce the indoor particle concentration to about 2×10^4 pc/L. If η is increased by 1.05 times to 0.995, while the nominal flow rate, *i.e.* n, remains the same, the corresponding CADR increase by 1.05 times. In this case, the CADR value can be considered unchanged. However, the stead-state concentration will not be changed by 1.05 times or remain unchanged. Instead, it is reduced by 80%. And the steady-state concentration approaches 4000 pc/L. This means under the condition of

Table 5.1 Concentration of the atmospheric dust in polluted situation (≥0.5 μm)

Type of polluted air	Concentration (pc/L)
US industrial air	3.5×10^5
Pollution	5.6×10^5
With photochemical smog	10^6
Seriously polluted	1.75×10^6
Pollution	10^6
With smog	2×10^6
Common industrial air in China	$(2–3) \times 10^5$

Note Concentration of microorganism in atmospheric dust is between 1 and 1000 pc/L

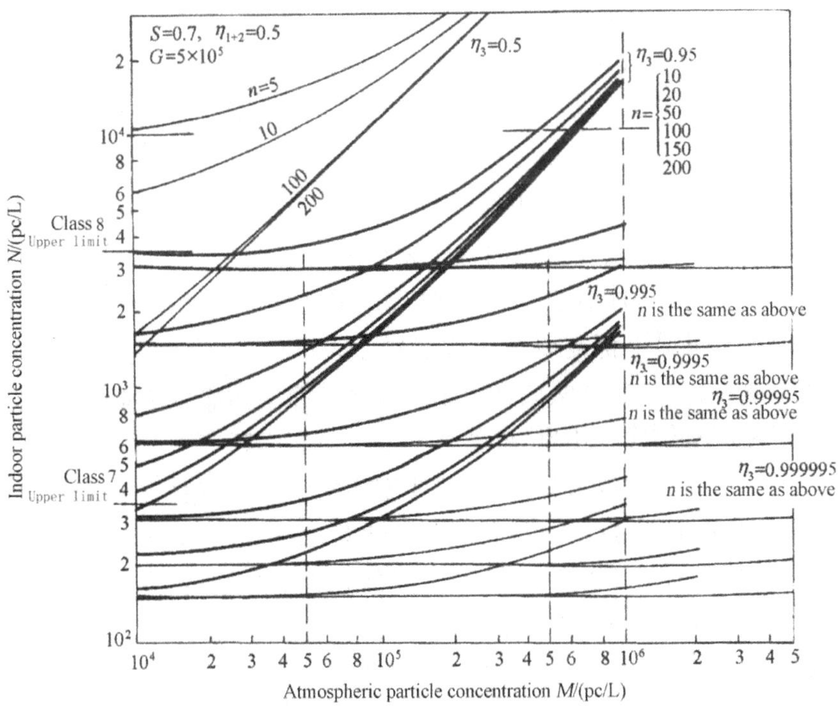

Fig. 5.3 One of the characteristic curves in at-rest cleanroom

uniform distribution, even when the CADR is equivalent, the purification capability may be significantly different.

5.3.3 *Summary*

In short, the concept of CADR cannot be used to explain the second situation as well. Taking the second situation as an example, the purification performance with the increased flow rate by 10 times (i.e. CADR increased by 10 times) is not as good as that with increased value of η by 1.05 times (i.e. CADR increased by 1.5 times). For the second situation, the purification performance with the increased value of η from 0.6 to 0.992 (by 1.24 times) is not as good as that with the increased value of η from 4 to 4.96 (by 1.24 times). For the latter case, since the value of Ψ will be reduced, the corresponding purification performance will be improved by more than 1.24 times.

5.4 Operational Property in Conditions with Complete Indoor Self-circulation

5.4.1 *General Formula for Steady-State Particle Concentration*

The condition with indoor self-circulation is referred as the third situation before. When the particle indoors is not uniformly distributed and there is no outdoor air ($s = 1$) in Eq. (5.15), the average steady-state concentration can be obtained from Eq. (5.16).

$$N = \psi \frac{60G \times 10^{-3}}{n\eta} \quad (5.16)$$

5.4.2 *Relationship Between Efficiency, Air Change Rate and Particle Concentration*

For rooms with air purifier, the relationship between Ψ and n is much complex than that in cleanroom. For the latter case, Ψ may be larger than 2 when $n < 5$, while Ψ may be smaller than 1 when $n > 80$. When n is reduced from 6 to 3 h^{-1}, the indoor particle concentration will be increased by more than 1 times because Ψ will increase from 1.6 to 3.5. The total N value will increase by 2 times. For air purifier, the difference of the result will be much significant. In short, the purification capability is not proportional to the values of n and η of this air purifier.

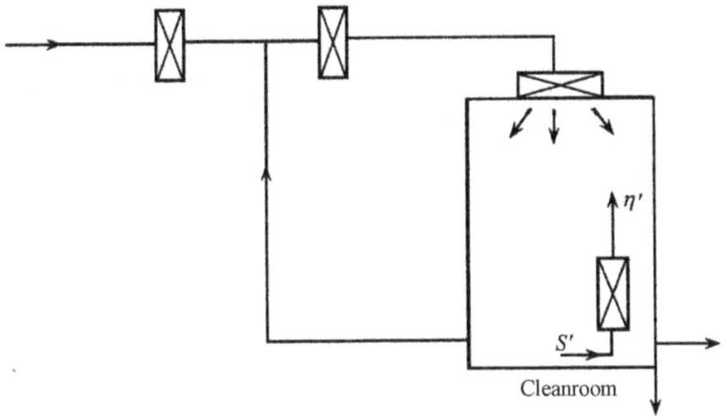

Fig. 5.4 Schematic diagram of the room with local air purifier

5.4.3 Summary

In summary, the concept of CADR cannot be used to explain the third situation as well. For the common flow rate rage of portable air purifier, when the flow rate is reduced by a half, the average particle concentration may be increased by 4 times. Therefore, the purification capability of the air purifier may be worse even though it looks small and exquisite.

5.5 Operational Property in Conditions Supplementary Function

5.5.1 General Formula for Steady-State Particle Concentration

When air purifier is placed in the room where there is central ventilation or air conditioning system, it is the fourth situation mentioned before. It is shown in Fig. 5.4.

In Fig. 5.4, there is both centralized system and air purifier. The steady-state particle concentration in this case becomes [7]:

$$N = \frac{60\mathrm{G} \times 10^{-3} + \mathrm{Mn}(1 - \mathrm{s})(1 - \eta_{\mathrm{n}})}{\mathrm{n}[(1 + \eta' \mathrm{s}') - \mathrm{s}(1 - \eta_{\mathrm{t}})]} \tag{5.17}$$

where

s' is the ratio of the return air in the local air purifier, which is the percentage of the self-circulated air in the total flow rate;
η' is the total efficiency of various air filters in the local air purifier.

Suppose there is air filter with performance equivalent to or better than sub-HEPA filter in air purifier, $\eta' \approx 1$. Equation (5.17) can be rewritten as:

$$N = \frac{60G \times 10^{-3} + Mn(1 - s)(1 - \eta_n)}{n[(1 + s') - s(1 - \eta_t)]} \tag{5.18}$$

The difference between Eq. (5.18) and previous equations is the denominator, where "1" is changed into "1 + s'". This can be considered that the air change rate of the room is improved when local air purifier is placed indoors.

5.5.2 Noteworthy Points

If $s(1 - \eta_t)$ is much smaller than 1, Eq. (5.18) can be simplified as Eq. (5.19).

$$N = \frac{60\mathrm{G} \times 10^{-3} + \mathrm{Mn}(1 - \mathrm{s})(1 - \eta_\mathrm{n})}{\mathrm{n}(1 + \mathrm{s}')} \tag{5.19}$$

With this equation, the influence of the local air purifier on the particle concentration can be found. Only a very small proportion of the air change rate is increased.

The feature of this situation is the same as that of the second situation, which will not be analyzed further.

It should be noted that the air distribution created by the existing central air conditioning system may be easily disturbed by setting of air purifier. Especially in applications where unidirectional airflow is needed such as ICU, the whole purification performance may be influenced. In some cases, the side effect may be created. This must be paid special attention to. In GB50736-2012 "*Design code for heating ventilation and air conditioning of civil buildings*", it has been pointed out in Item 7.4.7 that "there should be no other air distribution in the air conditioning region". This should also be required for the cleaning region.

References

1. Z. Xu, Y. Zhang, G. Cao, X. Feng, H. Pan, Study on performance evaluation index of air purifier-discussion on several problems of air purifier. J. HV&AC **47**(8), 2–6 (2017)
2. Z. Xu, *Fundamentals of Air Cleaning Technology*, 1st edn. (China Architecture & Building Press, Beijing, 1983)

3. J. Shen, Y. Liu, J. Yan, Correct understanding of medical environment control technology. J. HV&AC **46**(6), 73–78 (2016)
4. Z. Xu, *Fundamentals of Air Cleaning Technology*, 4th edn. (Science Press, Beijing, 2014), p. 394
5. Z. Xu, *Fundamentals of Air Cleaning Technology*, 4th edn. (Science Press, Beijing, 2014), p. 66
6. Z. Xu, *Fundamentals of Air Cleaning Technology*, 4th edn. (Science Press, Beijing, 2014), p. 410
7. Z. Xu, *Fundamentals of Air Cleaning Technology*, 4th edn. (Science Press, Beijing, 2014), p. 389

Chapter 6
Application Property of Air Purifier

The purpose for application of air purifier indoors is to reduce the indoor particle concentration and then the particle concentration becomes stable within a certain period. This period is called the self-purification time. The less it is, the better it will be. In order to meet the above requirements, what is the suitable area for one air purifier? Is that any difference between different applications? Therefore, this chapter focuses on the concerned problem for application of air purifier [1]: the self-purification time to reach the steady state concentration indoors (there will also be a process of small fluctuation after the steady state concentration is basically reached); the applicable situation and the applicable area (introduced in Chap. 8), etc. As for the fundamental to these problems, i.e., the value of the steady-state particle concentration, it is influenced by many factors, which will be discussed in Chap. 7.

6.1 Self-purification Time

6.1.1 Theoretical Calculation

People care most is the two conditions for independent operation of air purifier among the above-mentioned four conditions in Chap. 5, that is:

(1) The self-circulation condition when outdoor air is delivered into the air purifier, which is the condition No. 2.
(2) The self-circulation condition when there is particle source indoors and there is no outdoor air delivered into the air purifier (there is also no dedicated outdoor air system), which is the condition No. 3.

Z. Xu, *Air Purifier*, https://doi.org/10.1007/978-981-13-2532-8_6

The condition No. 1 is the condition in the test chamber. In theory, when $t \to \infty$, $N_t \to 0$. In this case, the steady-state concentration and its self-purification time are not so important.

The condition No. 4 is mainly concerned with the system, which is similar to the condition No. 2 in Chap. 5 so it is not included in the discussion here.

The transient particle concentration for condition No. 2 where the outdoor air is introduced has been given in Eq. (5.4). It is re-written here again:

$$N_t = \frac{60G \times 10^{-3} + Mn(1-s)(1-\eta_n)}{n[1-s(1-\eta_r)]} \cdot \left\{1 - \left[1 - \frac{N_0 n[1-s(1-\eta_r)]}{60G \times 10^{-3} + Mn(1-s)(1-\eta_n)}\right] e^{-nt[1-s(1-\eta_r)]/60}\right\} \quad (6.1)$$

where

η_n is the total efficiency of air filters on the pipeline of outdoor air;
η_r is the total efficiency of air filters on the pipeline of return air;
G is the uniformly-distributed particle generation rate by indoor particle generation source, pc/(m^3 min); n is the air change rate, h^{-1};
N_0 is the initial particle concentration indoors, pc/L;
N_t is the transient particle concentration indoors at time t, pc/L;
M is the concentration of atmospheric dust, pc/L;
s is the ratio of return air;
t is the time, min;

When only air purifier is used, $\eta_n = \eta_r = \eta$. The parameter η is the total efficiency of air purifier. For the condition No. 3 where there is no introduction of outdoor air, Eq. (6.1) can be simplified as Eq. (6.2).

$$N_t = \frac{60G \times 10^{-3}}{n\eta} - \frac{60G \times 10^{-3}}{n\eta} e^{-n\eta t/60} + N_0 e^{-n\eta t/60} \quad (6.2)$$

Since the first item outside of the bracket {} in Eq. (6.1) and the first item in Eq. (6.2) are steady-state concentration N, the simplified equation can be obtained for condition No. 2 or No. 3:

$$N_t = N\left(1 - e^{-n\eta t/60}\right) + N_0 e^{-n\eta t/60} \quad (6.3)$$

Since values of N_t and t are unknown, the self-purification time t cannot be calculated. Till now there is no good solution to this problem abroad. This problem has already been solved by author with the graphic method in 1975 in China [2].

With Eq. (6.3), we can derive:

$$\frac{N_t}{N} = \left(1 - e^{-n\eta t/60}\right) + \frac{N_0}{N} e^{-n\eta t/60}$$
$$\frac{N_t}{N} - 1 = \left(\frac{N_0}{N} - 1\right) e^{-n\eta t/60} \quad (6.4)$$

It is reasonable to assume $N_t = (1.01, 1.001, \ldots, 1.0001)N$. For practical point of view, it is enough to assume $N_t = 1.01\,N$. Similarly, we can assume $N_0 = (10, 20, \ldots)N$. In terms of smog removal, $N_0/N = 10$ is enough for the requirement. This is because even when the outdoor $PM_{2.5}$ concentration reaches the extreme value 700 μg/m^3 (The initial indoor concentration N_0 can usually reach 75% of the outdoor concentration, which is shown in Chap. 8. For safety reason, the outdoor concentration can be used directly for N_0), if the indoor $PM_{2.5}$ concentration is reduced to the steady-state value 70 μg/m^3, which is 1/10 of the outdoor concentration, it is already within the upper limit 75 μg/m^3 of the second level and it belongs to "good" condition. Usually, N_0 could be considered as 350 μg/m^3 (This is the situation in Beijing. The initial indoor concentration in Shanghai is lower than this value.). Especially when there is only indoor particle generation source (including the infiltration of particles) and when self-purification with air purifier is used, 1/10 of the outdoor concentration means that N should be 35 μg/m^3, which has reached the standard of the first level. Of course, the precondition is that the air puffier is able to reduce the indoor concentration to 1/10 of N_0. Of course, one sixth of 350 μg/m^3 is already within the standard of second level. This will be discussed in Chap. 7.

Equation (6.4) can be re-written as:

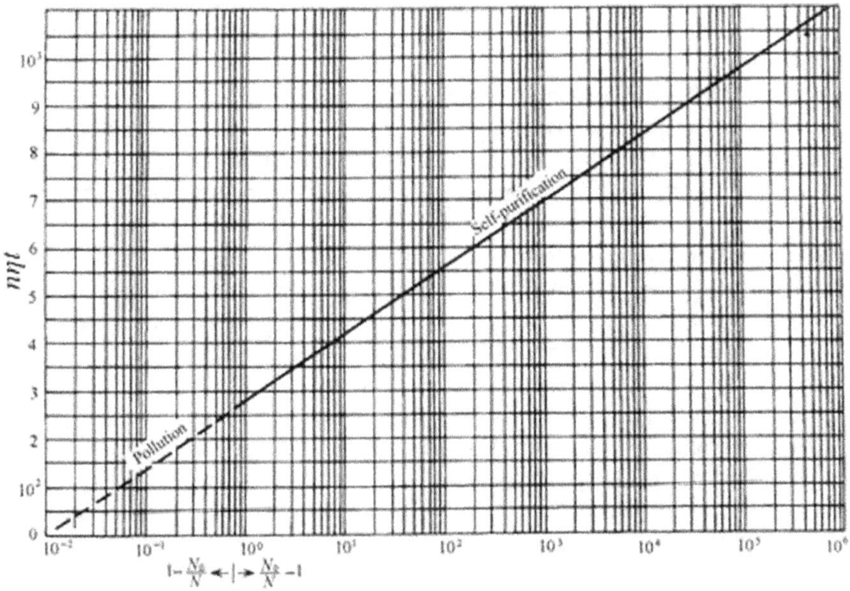

Fig. 6.1 Calculation figure for self-purification time

$$0.01 = \left(\frac{N_0}{N} - 1\right) e^{-\mathrm{n\eta t}/60}$$

$$\mathrm{n\eta t} = 60\left[\ln\left(\frac{N_0}{N} - 1\right) - \ln 0.01\right] \quad (6.5)$$

With this formula, graphic calculations figure can be plotted, which is shown in Fig. 6.1. $n\eta$ can also be expressed with the decay coefficient k.

In this figure, N_0 is the initial particle concentration indoors, pc/L; N is the steady-state particle concentration indoors needed (it should be realized), pc/L; n is the air change rate (or it should be calculated), h^{-1}; t is the self-purification time needed (or it is the practical value or designed value), min.

For the following discussions, except for special explanation, the particle counting efficiency with diameter ≥ 0.5 μm and the particle counting concentration will be used.

Based on the above calculation figure, the following results of N_0/N can be obtained for different reachable groups of steady-state concentrations N (Note: it is not the transient concentration N_t), which is shown in Table 6.1.

It is shown in the above table that suppose the value of N_0 is the same under the condition of $n = 6$ h^{-1} and $\eta = 0.9$, the steady-state concentration N, which is equivalent to one sixth of N_0, can be reached because of large amount of particle generation indoors. This means the value of N is relative large and it costs 69.4 min to be stable. Or when N is constant but N_0 becomes large, N_0/N reaches 10, so the time to reach the same concentration value N is much longer.

It is shown from the above results that when the particle counting efficiency η increases from 0.9 to 0.9999, HEPA filter will be used to replace high-medium efficiency air filter (only barrier-type air filter has such broad spectrum of efficiency). But the self-purification time is not shortened obviously (about 10%), and the pressure drop is increased greatly. Instead, when air purifier with larger flow rate is used and the air change rate n is increased from 6 to 8 h^{-1}, t will be shortened by about 20%. When the air change rate n is 10 h^{-1}, t will be shortened by about 40%, but it should be noted that it is difficult to reach 10 h^{-1} for air purifier.

The difference is the ordinate, where the expression nt for cleanroom is changed into $n\eta t$. For cleanroom where three stages of air filtration system are used, $\eta \approx 1$.

When the value N is fixed, N_0/N is obtained. The value of $n\eta t$ can be obtained from Fig. 6.1 by the intersection point between the straight line for $N_0/N - 1$ and the oblique line. So we get $\mathrm{t} = n\eta t/n\eta$.

6.1.2 Measured Time

The calculated value of air-purification time based on the calculation graph agrees well with the measured results. Some examples will be listed below:

Table 6.1 Calculation of self-purification time

$N_0/N = 20$, look at Fig. 6.1	n/h^{-1} (min^{-1})	η	t/min	$N_0/N = 10$, look at Fig. 6.1	n/h^{-1} (min^{-1})	t/min	$N_0/N = 6$, look at Fig. 6.1	n/h^{-1} (min^{-1})	t/min
$n\eta t \approx 45$	6(0.1)	0.9	83.9	$n\eta t \approx 40$	6(0.1)	75.6	$n\eta t \approx 375$	6(0.1)	69.4
		0.99	76.3			68.7			63.1
		0.9999	75.5			68.0			62.5
	8(0.13)	0.9	60.9		8(0.13)	56.7		8(0.13)	51.8
		0.99	57.2			51.5			47.0
		0.9999	56.6			51.0			46.6
	10(0.167)	0.9	50.3		10(0.167)	45.3		10(0.167)	41.4
		0.99	45.8			41.2			37.7
		0.9999	45.3			40.8			37.3
	4(0.065)	0.9	125.8		4(0.065)	113.3		4(0.065)	103.6
		0.99	114.4			103.0			94.2
		0.9999	113.3			102.0			93.3
	2(0.033)	0.9	251.7		2(0.033)	226.7		2(0.033)	207.2
		0.99	228.8			206.1			188.4
		0.9999	226.5			204.0			186.5

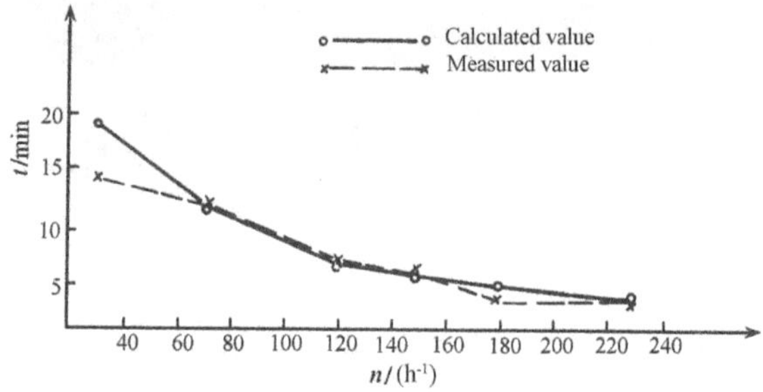

Fig. 6.2 Relationship between self-purification time and the air change rate

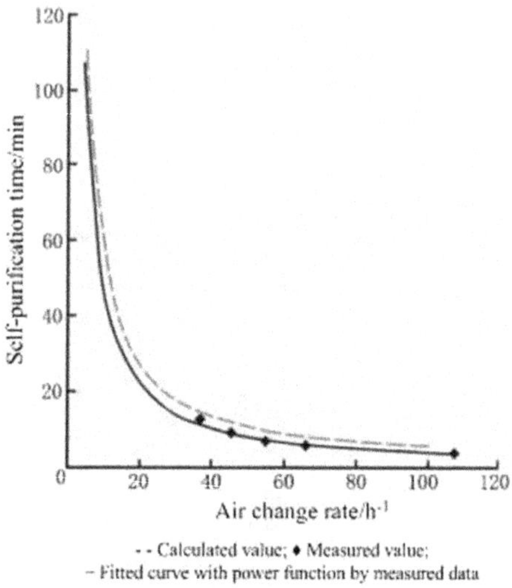

Fig. 6.3 Self-purification time from Class 7 to Class 5 for newly-built sterile injection workshop with small volume

(1) Comparison with measured data by Oshitari Laboratories, Inc. in Japan [3]

The dashed line in Fig. 6.2 shows the relationship between self-purification time and the air change rate given by Oshitari Laboratories, Inc. in Japan. The curve by their data is almost consistent with the calculated curve (solid line in Fig. 6.2) with the calculation graph in Fig. 6.1.

(2) Comparison with measured data in pharmaceutical workshop [4]

Figure 6.3 shows the comparison between the measured and the calculated results for GMP pharmaceutical workshop published in 2014 by Xin Feng et al. It is shown that two curves agree very well.

(3) Comparison with measured data for air purifier [5]

During the severe haze condition, the purification performances of two domestic air purifiers with HEPA filter, whose prices differ a lot, were measured in residence of Beijing by Li et al. TSI9306-V2 handheld particle counter was used to measure the airborne particle concentration. The sampling rate was 2.83 L/min. Three times of

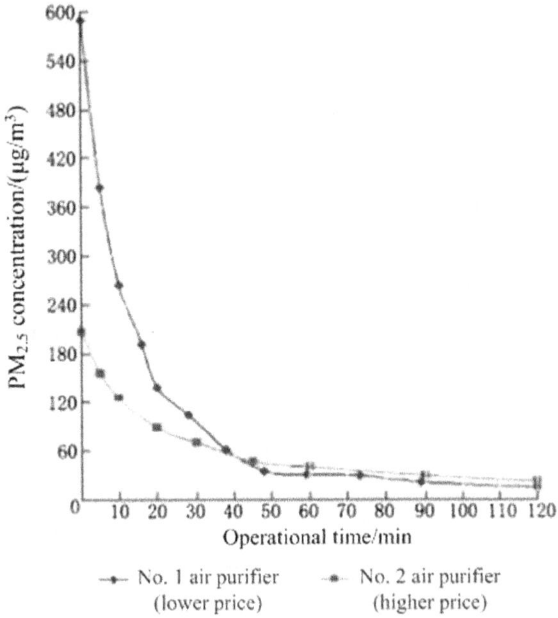

Fig. 6.4 Variation of indoor $PM_{2.5}$ concentration after turn on of two air purifiers

sampling were performed at each sampling position. After the air purifier was turned on, two times of sampling were performed at each sampling position (because the variation of concentration was too rapid). Meanwhile, QD10 $PM_{2.5}$ sampler was used to measure the gravimetric concentration of $PM_{2.5}$. Anemometer was used to measure the air velocity at the outlet.

(a) The air change rate provided by No. 1 air purifier was 5.2 h^{-1}. Coarse, HEPA filters and activated carbon layer were installed. Air purifier was installed near the exterior window at the height of 0.8 m above the floor. The outer shape was cylindrical. Air was returned through the upper side. Air was supplied at the lower side. Air purifier was turned on when the particle concentration for particle diameter ≥ 0.3 μm was 1,003,821,000 pc/m^3 (i.e., $>10^6$ pc/L). The self-purification curve is shown as the upper curve in Fig. 6.4. After 90 min of operation, the particle concentration was almost stable.

It is shown in Fig. 6.4 that steady state was achieved after 90–120 min. At 120 min, the particle concentration for diameter ≥ 0.3 μm was given in the literature, which was 33,390,111 pc/m^3. The detailed data at 90 min was not provided in the paper. Suppose that the data at 120 min was used instead, it is shown in Fig. 6.4 that the difference between them should be very small. Then $N_0/N \approx 1{,}003{,}821{,}000/33{,}390{,}111 = 30$. Check Fig. 6.1, we can obtain $n\eta t \approx 470$. With HEPA filter, $\eta \approx 1$. So we can get: $t = 470/5.2 = 90.4$ min. This means it takes more than 90 min to reach the steady state. This is quite consistent with the conclusion by this literature. Even though the difference reaches about 10%, the agreement is still quite good.

(b) The air change rate provided by No. 2 air purifier was 5.3 h^{-1}. Coarse, HEPA filters and activated carbon layer were also installed. This air purifier was more expensive than No. 1 air purifier by ten times. The installation position was the same as that of No. 1. The outer shape was rectangular. Air was return from the front face. Air was supplied at the upper side.

It was shown from the experiment that indoor particle concentration reached 25 $\mu g/m^3$ at 80–90 min with the No. 2 air purifier (the initial particle concentration was 210 $\mu g/m^3$ in the figure). Then $N_0/N = 8.4$. According to Fig. 6.1, we can obtain $n\eta t \approx 400$. Since $\eta \approx 1$ and $n = 5.3\ h^{-1}$, we can get: $t = 400/5.3 = 76$ min. This is quite close to the measured value. The positioning of sampling points has relative large influence on measurement results. Therefore, before the steady state is reached, the times needed to reach N_0/N differ a lot, which can also be found from the data in this literature.

The air change rate n during experiment is slightly smaller than 6 h^{-1} in Table 6.1. With three stages of air filtration system, η is slightly larger than 0.9999 in Table 6.1. It is shown in Fig. 6.4 that value of N between 60 and 70 min is almost the same. N_0 is about 11 times of N. In Table 6.1, the concentration at 68 min is about 10 times of N_0. They are basically equivalent.

When the expected reduction ratio from the original concentration is known by user, as long as the value of t is obtained by test, the applicable area can be obtained. This method is simple. And it is easy for user to know the degree of purification. If the penetration coefficient of outdoor particle, indoor concentration and natural deposition coefficient are needed for direct calculation, the method will be much complex.

6.2 Applicable Situation

6.2.1 Air Purifier with Indoor Particle Generation and Self-circulation

For the air purifier used in the above literature, after operation time of 90–120 min, the indoor particle concentration for diameter $\geq$0.3 μm was reduced to 33 390,111 pc/m^3 (>3.3 × 10^4 pc/L [4]), and that for diameter $\geq$0.5 μm was about 1 × 10^4 pc/L. The ratio of particle concentration between $\geq$0.5 and $\geq$0.3 μm is 1/3. Although they are less than the upper limit of ISO Class 8.5 (Class 300 000 based on US FS209) and there was no outdoor air introduced into the air purifier, it took 1.5–2 h. So it is impossible to play its role mainly in cleanroom.

6.2.2 Air Purifier with Indoor Particle Generation and Introduction of Partial Outdoor Air

For ordinary weather, $M = 2 \times 10^5$ pc/L. When the fresh air ratio is assume 0.3 and the efficiency of air purifier is $\eta = 0.995$, the air change rate should be $n = 10\ h^{-1}$ so that the upper limit of ISO Class 8 can be reached, or $n = 20\ h^{-1}$ so that the intermediate value of ISO Class 8 can be reached. It is difficult to set the local cleaning device-air purifier with such large amount of flow rate. It is also impractical. If η is small, even through the flow rate is increased by 10 times, the purification ability can only be increased by 2–3 times. It is difficult to explain this by the relationship in "CADR" equation. But when η is increased to 0.99999 (which is equivalent to 0.9999 with 0.3 μm DOP), the performance with the air change rate 10 h^{-1} is equivalent with the former case under the air change rate 200 h^{-1}. But It is shown in Fig. 5.3 that it is unnecessary to set η more than 0.99999. Meanwhile, the problem of temperature and humidity control cannot be solved by air purifier. Even with $\eta = 0.99999$ and $n = 10$ in the ordinary weather without smog, indoor particle concentration can only reach the upper limit of ISO class 8. The flow rate of the air purifier will be reduced gradually to 50%, but the flow rate for the cleanroom is not allowed to be lower than the standard value (through adjustment).

It should be noted that when the theory of expanded mainstream area proposed by author is used for calculation [4, 6, 7], the higher air cleanliness level can be obtained in cleanroom. But for the room with application of air purifier, the theory of expanded mainstream area cannot be used for calculation.

6.2.3 Air Purifier with Supplementary Function to Existing System

In the special vortex area or the unreachable area by airflow, or there is inherent shortcoming of the existing system, air purifier can be considered as the supplementary role. Examples include the edge of the place with relative large area, local enclosure area, etc. For application where the direction of airflow should be kept consistent with the deposition direction of particles, it is infeasible to add local purification equipment. It has been pointed out in Chinese national standard "*Design code for heating ventilation and air conditioning of civil buildings*" that there should be no other air distribution inside the air conditioning area.

In terms of health care facilities, only standards from U.S., China and Japan mention the indoor self-circulation equipment. Others only mention ventilation and air conditioning system, while self-circulation equipment is not mentioned. In ANSI/ASHRAE/ASHE-170-2013 "*Ventilation of Health Care Facilities*", 2013 (Table 6.2) the indoor self-circulation was mentioned and the requirement was the most stringent, which is shown in Fig. 6.2 [8].

Table 6.2 Specifications on indoor self-circulation equipment in ASHRAE170 standard

Function of space	Setting of indoor self-purification equipment
Surgery and critical area	
Class B and C operating room	No
Operating/surgical cystoscopic rooms	No
Delivery room (Caesarean)	No
Substerile service area	No
Recovery room	No
Critical and intensive care	No
Intermediate care	No requirement
Wound intensive care (burn unit)	No
Newborn intensive care	No
Treatment room	No requirement
Trauma room (crisis or shock)	No
Medical/anesthesia gas storage	No requirement
Laser eye room	No
ER waiting rooms	No
Triage	No requirement
ER decontamination	No
Radiology waiting room	No requirement
Class A operating/procedure room	No
ER check/treatment room	No requirement
Inpatient nursing	
Patient room	No requirement
Nourishment area or room	No requirement
Toilet room	No
Newborn nursery suite	No
Protective environment room (abbreviated as PE room)	No
Isolation room with airborne contamination (abbreviated as all room)	No
Combination all/PE room	No
Anteroom for isolation room with airborne contamination (abbreviated as all anteroom)	No
Protective environment anteroom	No
Combination all/PE anteroom	No
Labor/delivery/recovery/postpartum (LDRP)	No requirement
Labor/delivery/recovery (LDR)	No requirement
Patient corridor	No requirement

(continued)

Table 6.2 (continued)

Function of space	Setting of indoor self-purification equipment
Nursing facility	
Resident room	No requirement
Resident gathering/activity/dining	No requirement
Corridor for resident unit	No requirement
Physical therapy	No requirement
Occupational therapy	No requirement
Bathing room	No
Radiology	
X-ray (diagnostic and treatment)	No requirement
X-ray (surgery/critical care and catheterization)	No
Darkroom	No
Diagnostic and treatment	
Bronchoscopy, sputum collection, and pentamidine administration	No
Laboratory, general	No requirement
Laboratory, bacteriology	No requirement
Laboratory, biochemistry	No requirement
Laboratory, cytology	No requirement
Laboratory, glasswashing	No requirement
Laboratory, histology	No requirement
Laboratory, microbiology	No requirement
Laboratory, nuclear medicine	No requirement
Laboratory, pathology	No requirement
Laboratory, serology	No requirement
Laboratory, sterilizing	No requirement
Laboratory, media transfer	No
Nonrefrigerated body-holding room	No
Autopsy room	No
Pharmacy	No requirement
Examination room	No requirement
Medication room	No requirement
Gastrointestinal endoscopy procedure room	No
Endoscope cleaning	No
Treatment room	No requirement
Hydrotherapy	No requirement
Physical therapy	No requirement

(continued)

Table 6.2 (continued)

Function of space	Setting of indoor self-purification equipment
Sterilizing	
Sterilizer equipment room	No
Central medical and surgical supply	
Soiled or decontamination room	No
Clean workroom	No
Sterile storage	No requirement
Service	
Food preparation center	No
Ware washing	No
Dietary storage	No
Laundry, general	No
Soiled linen (sorting and storage)	No
Clean linen storage	No requirement
Linen and trash chute room	No
Bedpan room	No
Bathroom	No
Janitor's closet	No
Support space	
Soiled workroom or soiled holding	No
Clean workroom or clean holding	No requirement
Hazardous material storage	No

In Japanese standard relate to hospital, it is required that indoor self-purification equipment is not allowed in some rooms, while it can be used in some rooms with provisional conditions. For most of the ordinary rooms, self-purification equipment can be used.

It is specified in U.S. standard that the indoor self-circulation equipment in the table does not include heating and cooling coil. Because of the difficulty of cleaning and accumulation of pollutant, indoor self-circulation equipment is not allowed to be used in region with label "No". But self-circulation equipment with HEPA filter should be used as supplementary method temporarily in existing facilities. This is the same as the fourth operational condition of air purifier in this book. It is obvious that the indoor self-circulation unit here mainly means the air purifier.

In Chinese national standard GB50333-2013 "*Architectural technical code for hospital clean operating department*", the requirement is also not very strict. Except the class I–III and negative pressure clean operating rooms, self-circulation equipment can be used in other clean facilities. But as mentioned before, it will not be used until it is urgently needed.

6.3 Conclusion

In summary, air purifier with a certain filtration efficiency and flow rate is feasible for general indoor environment and part of the medical space. But there is no fundamental basis for the idea of using air purifier instead of clean air conditioning system in cleanroom. When it is used in rooms with existing air conditioning system, it is also basically infeasible for the space where the direction of airflow is required.

References

1. Z. Xu, Y. Zhang, G. Cao, Y. Zhang, W. Niu, Applicable situation and cleanliness recovery time of air cleaners—Discussion on several problems about air cleaners. J. HV&AC **47**(8), 7–10 (2017)
2. Z. Xu, *Fundamentals of Air Cleaning Technology*, 1st edn. (China Architecture & Building Press, Beijing, 1983)
3. Oshitari Laboratories Inc., *Test Report of Cleanroom for Experiment* (1974)
4. X. Feng et al., Field tests and analysis for air change rate needed for GMP Grade B area-the background area for manufacturing sterile medicinal products. J. HV&AC **44**(2), 84–88+132 (2014)
5. Z. Li, K. Xing, L. Yang, X. Qiu, Test and analysis on anti-haze effect of domestic air purifiers under severe haze condition. J. HV&AC **46**(6), 1–4+32 (2016)
6. Z. Xu, *Fundamentals of Air Cleaning Technology*, 4th edn. (Science Press, Beijing, 2014)
7. Z. Xu, *Design, Operation and GMP Accreditation on Pharmaceutical Factory*, 2nd edn (Tongji University Press, 2011)
8. Compiling Group for "Architectural technical code for hospital clean operating Department," "Translation Selected Referece Material for Foreign Construction Standards on Medical Facilities," (2012)

Chapter 7
Calculation on Pollution Load of Air Purifier

7.1 Gap on External Window Is Main Route for Entrance of Pollutant

7.1.1 Outline

The pollution load of air purifier mainly means the concentration of particles invaded into the room, i.e., the $PM_{2.5}$ concentration.

Several years ago, there were days with dust storms in Beijing. The sky was gloomy and gray. But there was no feeling of wind. When people went back home after one-day out, a layer of something could be seen on the table. The floor also seemed to be a little gloomy. Since the double layered window was used, how did the pollutant come indoors without the help of the wind?

Outdoor pollution sources will be the primary target of concern. Indoor pollution sources will be described in Chap. 8.

7.1.2 Basic Model of Apartment

At present, the general plan of the ordinary residential unit is roughly: entrance through the unit security door, or then enter into each family through the corridor, or then go upstairs through the staircase. For each family, there are usually two outer doors. One is the security door, and the other is the wooden door towards the living room. People can go into the balcony through the balcony door. There is double layered external window on the exterior side of the balcony. Or there is no interior door between the living room and the balcony, and people can enter into the balcony directly. Sometimes there is another external window towards the other direction.

Z. Xu, *Air Purifier*, https://doi.org/10.1007/978-981-13-2532-8_7

Fig. 7.1 Schematic diagram of the apartment

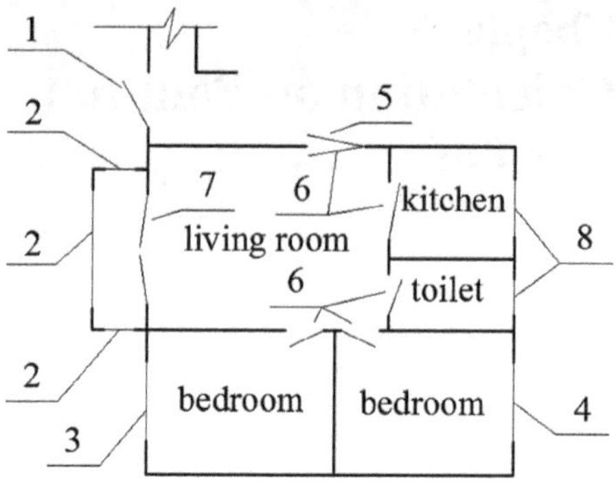

The bedroom is only connected with the outside through the external window. Figure 7.1 shows one schematic diagram of the apartment.

(1) External door of the unit; (2) Window at balcony; (3) Window in main bedroom; (4) Window in secondary bedroom; (5) Security door in living room; (6) Wooden interior door; (7) Door between living room and balcony; (8) Side window in kitchen.

7.2 Infiltration Rate of Particles by Entrance and Exit of Occupant Through Door

7.2.1 *Setting*

Suppose there is the door towards the outside directly. At the moment when people enter into the room against the opening direction of the door, the air velocity induced at the entrance is usually 0.08–0.15 m/s by experiment. This air velocity is smaller than 0.14–0.2 m/s which corresponds with the situation people enter into the room towards the opening direction of the door. It usually takes about 2 s for entrance of each people through the door [1]. The second wooden door is open towards indoors. The air velocity can be assumed 0.2 m/s. The cross-sectional area of occupant is about 1.7 m × 0.4 m. Therefore, the flow rate by entrance of occupant indoors is:

$$\Delta V = 1.7\ \text{m} \times 0.4\ \text{m} \times 0.2\ \text{m/s} \times 2\ \text{s} = 0.27\ \text{m}^3.$$

7.2.2 *Calculation*

Assume the particle concentration in the induced airflow is 10 times higher than the indoor concentration N. This can meet the requirements of heavy haze days.

Suppose the volume of the room (such as the small living room) is 1.5 m^2 × 2.6 m = 39 m^3, which can be set 40 m^3. There are two people entering into the room. Compared with the original particle concentration indoors, the amplification times of the indoor particle concentration by the induced airflow can be determined by the following equation, where the particle generation from occupant and the self-purification by the instantaneous airflow are not considered.

$$\frac{2\Delta V \times 10N + (V - 2\Delta V)N}{VN} = \frac{0.27 \times 2 \times 10N + (40 - 0.27 \times 2)N}{40N} = \frac{44.86}{40} = 1.12.$$

This means when there are two people walking into the room directly from the outside, the particle concentration in the small living room will be increased by 12% of the original concentration.

Since the number of entrance and exit in the general application is very small and the door of ordinary apartment or office will not be faced directly outwards (such as the unit apartment mentioned before), the induced particle pollutants from outside can be ignored completely. Therefore, for the residence and ordinary application, infiltration from external window becomes the main pollution source from outside.

7.3 Calculated Wind Velocity for Windward External Window

7.3.1 *Windward Wind Velocity*

In the building shown in Fig. 7.2 [2], positive pressure will be formed at the windward side, while negative pressure will be formed at the leeward side. For the ordinary room under natural condition, the pressure in the room will be lower than the pressure at the windward side. Even though the external window is sealed, infiltration through the window will occur. With the infiltration of outdoor air, the pressure in the room will be larger than the negative pressure at the leeward side. Indoor air will be driven outwards through the surface with negative pressure, as well as the window and door gap at non-positive pressure surface.

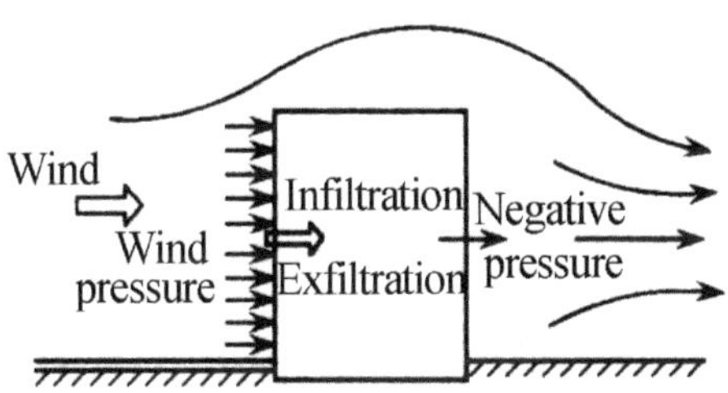

Fig. 7.2 Pressure caused by wind on wall surface

The pressure at the windward side is:

$$P_1 = C\frac{\rho u^2}{2} \text{ (Pa)} \quad (7.1)$$

where u is the calculated velocity at the windward side (m/s); According to GB50736-2012 "*Design code for heating ventilation and air conditioning of civil buildings*", the average wind velocity along the predominant wind direction in winter (except static wind) will be adopted. This will be explained in detail later.

ρ is the air density, and at normal temperature it is 1.2 kg/m^3, or the actual value is used;

C is the wind pressure coefficient, which can be set 0.9 on average.

7.3.2 Calculated Velocity at Windward Side

In Eq. (7.1), the calculated velocity u at the windward side is related to many factors, including air velocity at the windward side, building shape, dense level of buildings, layer height, orientation and temperature difference. For simplified calculation and safety, it is reasonable to set the average wind velocity along the predominant wind direction in winter for the value of u. If in some places where the wind velocity in summer is obviously larger than that in winter, the former value can also be used. The influence of u on P_1 is extremely large. When u is increased by one times, P_1 will be increased by four times.

The so-called average wind velocity along the predominant wind direction in winter means the average value of the average monthly wind velocity along the predominant wind direction during the most coldest months during continuous years (usually 30 years). The corresponding value in summer means the average value in the hottest three months. Table 7.1 shows the information [3].

In the table, there are 37 main cities. There are 25 cities where the wind velocity in winter is larger than that in summer. There is one city where the average wind velocities in winter and in summer are the same. There are 11 cities where the wind velocity in winter is smaller than that in summer. Therefore, the percentage of cities is 67.6% where winter velocity in winter is higher than that in summer. The percentage of cities is 29.7% where winter velocity in summer is higher than that in winter. Therefore, it is reasonable that the average wind velocity along the predominant wind direction in winter is used as the infiltrated calculated velocity at the windward side in Eq. (7.1) to calculate the wind pressure.

It is also shown in the table that the wind velocities in some western cities such as Urumchi, Chongqing and Lanzhou are much smaller than those coastal cities such as Dalian and Qingdao.

It is difficult to determine the influence of building height on the above velocity. For example, it has the great relationship with the dense level of buildings. The data in Table 7.1 were measured at the height of 10 m, which is the requirement of the

Table 7.1 Wind velocity in main cities

City	Beijing	Tianjin	Shanghai	Shijiazhuang	Taiyuan	Hohhot	Shenyang	Dalian
Predominant wind direction in winter	C N	C N	NW	C NNE	C N	C NNN	C NNE	NNE
Average wind velocity along predominant wind direction in winter/(m/s)	4.7	4.8	3.0	2	2.6	4.2	3.6	7.0
Predominant wind direction in summer	C SW	C S	SE	C S	C SN	C SW	SW	SSW
Average wind velocity along predominant wind direction in summer/(m/s)	3.0	2.4	2.6	2.6	3.0	3.4	3.5	4.6
City	Changchun	Jilin	Harbin	Nanjing	Suzhou	Hangzhou	Hefei	Fuzhou
Predominant wind direction in winter	WSW	C WSW	SW	C ENE	N	C N	C E	C NNW
Average wind velocity along predominant wind direction in winter/(m/s)	4.7	4.0	3.7	3.5	4.8	3.3	3.0	3.1
Predominant wind direction in summer	WSW	C SSE	SSV	C SSE	SE	S W	C SSW	SSE
Average wind velocity along predominant wind direction in summer/(m/s)	4.6	2.3	3.9	3	3.9	2.9	3.4	4.2
City	Xiamen	Nanchang	Jinan	Qingdao	Zhengzhou	Wuhan	Changsha	Guangzhou
Predominant wind direction in winter	ESE	NE	E	N	C NW	C NE	NNW	C NNE
Average wind velocity along predominant wind direction in winter/(m/s)	4.0	3.6	3.7	6.6	4.9	3.0	3.0	2.7
Predominant wind direction in summer	SSE	C WSW	SW	S	C S	C ENE	C NNW	C SSE
Average wind velocity along predominant wind direction in summer/(m/s)	3.4	3.1	3.6	4.6	2.8	2.3	1.7	2.3

(continued)

Table 7.1 (continued)

City	Shenzhen	Nanning	Haikou	Chongqing	Chengdu	Guiyang	Kunming	Lhasa
Predominant wind direction in winter	ENE	C E	ENE	CN NNE	C NE	ENE	C WSW	C ESE
Average wind velocity along predominant wind direction in winter/(m/s)	2.9	1.9	3.1	1.6	1.9	2.5	3.7	2.3
Predominant wind direction in summer	C ESE	C S	S	C ENE	C NNE	C SSW	C WSW	C SE
Average wind velocity along predominant wind direction in summer/(m/s)	2.7	2.6	2.7	1.1	2.0	3.0	2.6	2.7
City	Xi'an	Lanzhou	Xining	Yinchuan	Urumchi			
Predominant wind direction in winter	C NNE	C E	C SSE	C NNE	C SSW			
Average wind velocity along predominant wind direction in winter/(m/s)	2.5	1.7	3.2	2.2	2.0			
Predominant wind direction in summer	C ENE	C ESE	C SSE	C SSW	NNW			
Average wind velocity along predominant wind direction in summer/(m/s)	2.5	2.1	2.9	2.9	2.7			

Table 7.2 Wind velocity u on layers with different heights, m/s

Storey height	Shenyang	Tianjin	Beijing	Harbin	Dalian	Qingdao
3	3.6	3.6	4.3	4.6	6.4	7.0
6	3.6	3.8	4.5	4.5	6.5	7.0
30	2.1	2.9	4.5	5.1	7.5	7.0

meteorological station. According to the calculation analysis (shown in Table 7.2) [1], except in coastal cities where the calculated velocity can be increased for high-rise buildings, the influence of height in other regions can be omitted. Of course, this is a view.

Since the main concern here is the external window, the shaft effect between the internal staircases does not need to be considered.

7.4 Infiltration Airflow Through External Window by Differential Pressure

7.4.1 Calculation Equation

If the maximum calculated velocity in Beijing is 4.7 m/s along the north direction and the pressure indoors is 0 Pa, the pressure difference between indoors and outdoors formed by the wind pressure is:

$$P_1 - 0 = \Delta P_1 = 0.9 \times (4.7)^2 \times 1.2/2 = 12\ \text{Pa}$$

The maximum infiltration air velocity v through the gap regardless the facing direction can be given by the following equation [2]:

$$v = \varphi\sqrt{\frac{2\Delta P_1}{\rho}} \tag{7.2}$$

Since there is the resistance for the gap, $v < u$. The maximum infiltration air velocity is:

$$Q = 3600\ \mu F\sqrt{\frac{2\Delta P_1}{\rho}}\ \ (\text{m}^3/\text{h}) \tag{7.3}$$

where, μ is the coefficient of flow rate, which is usually 0.3–0.5. Since $\mu = \varepsilon\varphi$, where ε is the contraction coefficient, φ is the coefficient of air velocity and the theoretical

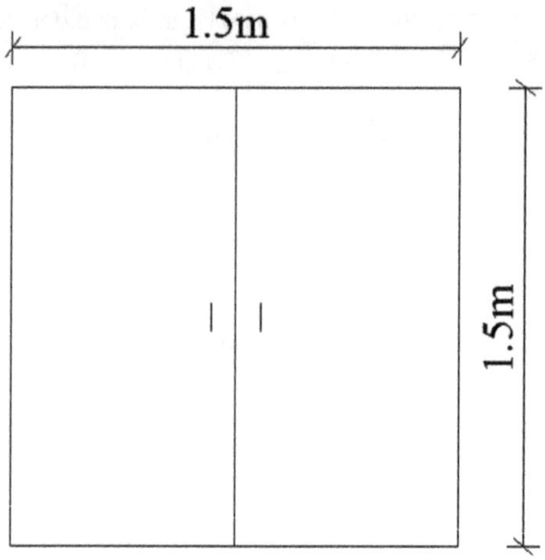

Fig. 7.3 Schematic diagram of gaps on external window of the bedroom

value is 0.82. According to author's measurement and analysis, the value of φ for the gap is very small. The minimum value can be 0.29 [4]. The value of φ for the gap can be assumed 0.6. So it is suggested $\mu = 0.3$.

F is the area of the gap, m^2.

7.4.2 Calculation on Infiltration Flow Rate and Window Gap Based on Standard External Window

It is difficult to know the exact area of the gap for the specific window.

Now this problem can be considered in another point of view.

In Chinese national standard GB/T7106-2008 "*Graduations and test methods of air permeability, watertightness, wind load resistance performance for building external windows and doors*", the air permeability performance for building external windows and doors is classified into different levels, which is shown in Table 7.3. This table was obtained when the total pressure difference ΔP was 10 Pa.

According to the specification in GB/T7106-2008, when the flow rate of leakage air for the unit gap length per hour is within some range, the different classification level can be obtained. The product with the minimum leakage flow rate by the current manufacturing process may be considered as the highest level.

If the double window with the size 1.5 m $\times$ 1.5 m is considered for each classification level, which is shown in Fig. 7.3, there are five gaps (only the gaps between the window frame and the sash are included) and the total length of gaps is 7.5 m. If the maximum leakage flow rate allowable for each classification level is used, the possible maximum gap width may be calculated, which is shown in Table 7.4.

Table 7.3 Classification for air permeability performance for building external windows and doors

Classification	1	2	3	4	5	6	7	8
Classification index through the unit gap length q_1/[m^3/(m·h)]	$4.0 \geq q_1 > 3.5$	$3.5 \geq q_1 > 3.0$	$3.0 \geq q_1 > 2.5$	$2.5 \geq q_1 > 2.0$	$2.0 \geq q_1 > 1.5$	$1.5 \geq q_1 > 1.0$	$1.0 \geq q_1 > 0.5$	$q_1 \leq 0.5$
Classification index through the unit area of gap q_2/[m^3/(m^2·h)]	$12 \geq q_2 > 10.5$	$10.5 \geq q_2 > 9.0$	$9.0 \geq q_2 > 7.5$	$7.5 \geq q_2 > 6.0$	$6.0 \geq q_2 > 4.5$	$4.5 \geq q_2 > 3.0$	$3.0 \geq q_2 > 1.5$	$q_2 \leq 1.5$

Table 7.4 Leakage flow rate of single window and possible maximum gap width (Pressure difference between indoors and outdoors $\Delta P_{\max} = 10$ Pa)

Classification level of window	1	2	3	4	5	6	7	8
Maximum allowable leakage flow rate of single window/(m^3/h) (Length of gaps is 7.5 m)	30	26.25	22.5	18.75	15	11.25	7.5	3.75
Possible maximum gap width (mm)	0.9	0.8	0.68	0.57	0.45	0.34	0.23	0.13

If ΔP_1 is not 10 Pa but $\alpha \Delta P_1$, the leakage flow rate becomes Q' and $Q' = \sqrt{\alpha} Q$. For example, when ΔP_1 is increased by one times, i.e., $\alpha = 2$, the leakage flow rate Q' will be $\sqrt{2} Q = 1.414Q$. If $\alpha = 0.05$, the leakage flow rate Q' will be $\sqrt{0.05} Q = 0.22Q$. In this case, when the pressure drop reduces from 10 to 0.5 Pa, the leakage flow rate is still more than 1/5 of the value with the calculated velocity of the maximum windward air velocity. For the Class 4 window with the above conditions, the leakage flow rate for each window under the pure wind pressure condition is about $0.22 \times 2.5 \times 7.5 = 4.125$ m^3/h. If the volume of the small bedroom is 10 m^2 $\times$ 2.6 m = 26 m^3 and there is a Class 4 window (total length of gaps is assumed 7.5 m), it should take at least 7 h so that the total infiltration flow rate is equal to the value of volume of this small bedroom.

7.5 Infiltration Airflow Through External Window by Temperature Difference

7.5.1 Thermal Pressure by Temperature Difference

It is already well-known for the infiltration of airflow through the gaps of external door and window by pressure difference. But when $PM_{2.5}$ is concerned, if there is no leakage by pressure difference, there should also be leakage by thermal pressure. The thermal pressure is created by the temperature difference. But it seems that not enough attention has been paid on the function of temperature difference.

When the weather is cold, this phenomenon of thermal pressure in kitchen is the most obvious. It is seen that hot air will be exhausted from the upper part of the screen window, but it is invisible at the lower part of the screen window. When hand is used, people will feel the infiltration of cold air from the outside.

It is also the phenomenon of thermal pressure for the wind bucket installed on the external window in the cottage in northern part of China in winter. When the stove is used for heating inside the room, indoor temperature will be higher and thermal pressure will be generated. The waste gas and particles from the stove will be exhausted from the wind bucket by the effect of thermal pressure. For natural balance, cold outdoor air will enter into the room through various gaps. This is why although window is not open, people do not feed stuffiness.

It is shown from Fig. 7.4 that the middle point "0" of the gap can be considered as the neutral point. Air flows inwards from half of the window, while air flows outwards from the other half of the window. Thermal pressure is created by the difference of air densities ρ with temperature difference at the inflow and outflow faces [2].

$$\Delta P_2 = gh\Delta\rho \quad \text{(Pa)} \tag{7.4}$$

For the upper part above the neutral point, the sign of ΔP_2 is "−". While for the lower part, the sign of ΔP_2 is "+".

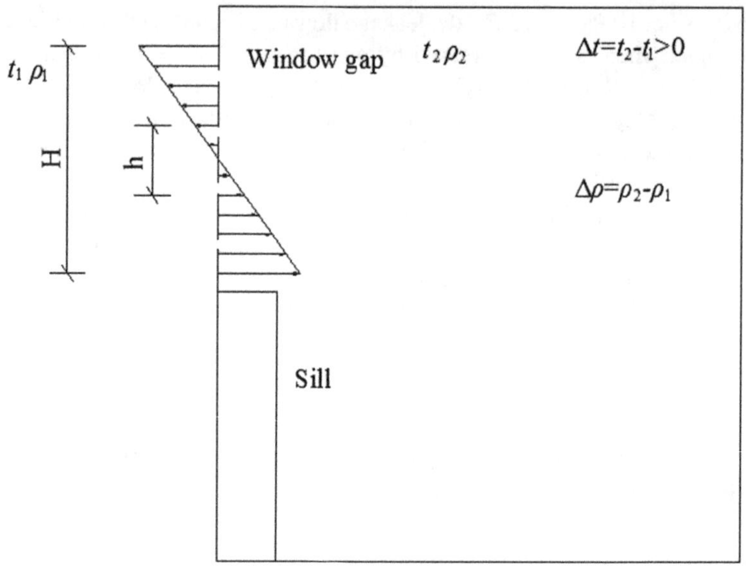

Fig. 7.4 Inward and outward airflow through door and window

Table 7.5 Density ρ of dry air/(kg/m^3)

t/°C	Atmospheric pressure/mm Hg			
	ρ			
	720	740	760	780
−10	1.271	1.307	1.342	1.377
0	1.225	1.259	1.293	1.727
10	1.182	1.215	1.247	1.280
20	1.141	1.173	1.205	1.237
30	1.104	1.134	1.165	1.196
40	1.069	1.098	1.128	1.158

Where, g is the gravitational acceleration, m/s^2;

h is the height difference between the center of the inflow surface and the center of the outflow surface. If the middle point is the neutral point, $h = H/2$, where H is the height of the gap, m;

$\Delta\rho$ is the difference of densities for the inflow and the outflow, kg/m^3. ρ is shown in Table 7.5 [5];

Because the temperatures of the inflow and the outflow are different, the density ρ of hot airflow (such as the exhaust airflow in winter) is small, and the density ρ of cold airflow (such as the infiltrated airflow in winter) is large.

It is shown in Table 7.5 that when the temperature difference is $\Delta t = 1$ °C under the normal atmospheric pressure, the difference of densities is $\Delta\rho = 0.0046$.

Based on Eq. (7.4), when $h = 1$ m and $\Delta t = 1$ °C, $\Delta P_2 = 9.8 \times 1 \times 0.0046 = 0.045$ Pa.

Because the value of thermal pressure is so small, it is likely to be ignored. But in Northern China, the temperature difference between indoors and outdoors in winter can usually reach 30 °C (−10–20 °C). In this case, $\Delta P_2 = 30 \times 0.045 = 1.35$ Pa. This pressure difference is equivalent to the wind pressure with the calculated wind velocity 1.5 m/s.

7.5.2 *Calculation on Infiltration Flow Rate by Thermal Pressure*

Suppose the infiltration with the wind pressure is not considered. When $\Delta t \geq 30$ °C, the density difference ρ between indoors and outdoors is relative large. The density ρ of hot outflow is small, so the velocity v is large. The density ρ of cold inflow is large, so the velocity v is small. Based on the principle of natural ventilation, when the middle point is still assumed for calculation on the area of the gap, the values of the inflow rate Q_2 and the outflow rate Q_2' with pure thermal pressure can be obtained with Eq. (7.3).

$$Q_2'' = 3600\,\frac{\mu F}{2}\sqrt{\frac{2\Delta P_2}{\rho_1}}\quad \left(\mathrm{m^3/h}\right) \tag{7.5}$$

$$Q_2'' = Q_2'$$

where, F is the total area of gaps, $F/2 = 3.75$ m × width of gap.

When the temperature of outdoor air is assumed −10 °C, $\rho_1 = 1.342$ kg/m^3. μ is assumed 0.3. The value of Q_2 can be seen in Table 7.6.

When there is completely no wind and the temperature difference between indoors and outdoors is 30 °C, for the Class 4 single window in the above-mentioned room with volume 26 m^3, it take about 8 h so that the total infiltration flow rate is equal to the volume of this room.

7.6 Infiltration Airflow Through External Window by Pressure-Temperature Differences

In the national standard GB50736, the calculation equation of infiltration flow rate under the combined effect of pressure and temperature differences is given, which considers various factors. But it is quite complex, and only empirical range is given for some parameters. In this section, the simplified calculation formula and the the-

Table 7.6 Leakage flow rate with temperature difference 30 °C

Classification level of window	1	2	3	4	5	6	7	8
Maximum leakage flow rate of single window/(m^3/h) (Total length of gaps is 7.5 m)	5.21	4.63	3.94	3.27	2.61	1.96	1.33	0.75

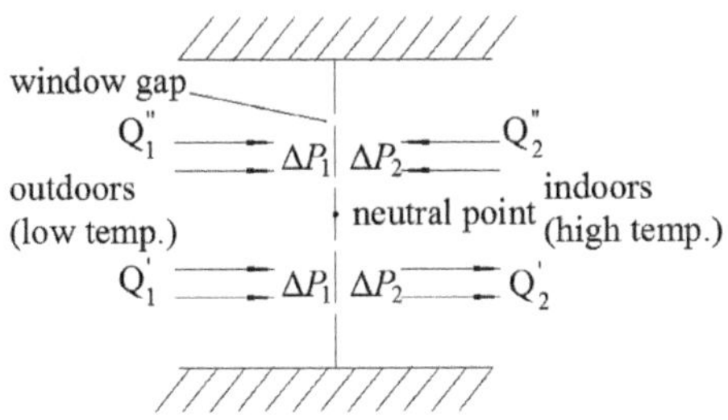

Fig. 7.5 Combined effect of the wind pressure and the thermal pressure

oretical value of infiltration flow rate under the combined effect of the wind pressure and the thermal pressure for pure external window will be given, which are provided as the supplementary method for the calculation method with natural ventilation.

As shown in Fig. 7.5, both the wind pressure and the thermal pressure exist on the cross section of window. But there are still several situations.

7.6.1 With Wind Pressure Only

Under the influence of the wind pressure, the infiltration flow rate can be calculated by Eq. (7.3), which can be simplified as Eq. (7.6).

$$Q = AF\sqrt{\Delta \mathrm{P}_1} \tag{7.6}$$

where A is the coefficient, and $A = 3600 \times 0.3 \times \sqrt{\frac{2}{1.342}}$. ΔP_1 is the pressure difference between indoors and outdoors with pure wind pressure. The value of 1.342 is the density ρ for inflow with temperature -10 °C. Of course, the actual value can be used.

It is shown that the infiltration flow rate when the windward pressure increases by one times is not as large as that when the area (or the length) of the gap increases by one times.

7.6.2 When Thermal Pressure Is Equal to Wind Pressure

This means $\Delta P_1 = \Delta P_2$.

It is obvious that there will no entrance and exit of airflow through the upper part of the window (The middle point of the window is assumed the neutral point). The inflow rate Q_1 at the lower part of the window

$$Q_1 = \frac{AF}{2}\sqrt{\Delta P_1 + \Delta P_2} = \frac{AF}{2}\sqrt{2\Delta P_1} = 0.707AF\sqrt{\Delta P_1} \tag{7.7}$$

It is obvious that $Q_1 < Q$.

7.6.3 Without Wind Pressure

This means $\Delta P_1 = 0$, and there is only the effect of thermal pressure.

It is obvious that there is only the infiltration airflow at the lower part of the window, which is symbolized as Q_2.

$$Q_2 = \frac{AF}{2}\sqrt{\Delta P_2} \tag{7.8}$$

Only when $\Delta P_2 > 4\Delta P_1$, the infiltration flow rate by pure thermal pressure is larger than that with the pure wind pressure ΔP_1.

7.6.4 When Thermal Pressure Is Smaller Than Wind Pressure

This means $\Delta P_2 < \Delta P_1$.

It is obvious that there is infiltration flow on all parts of the window, which is symbolized as Q_3.

$$Q_3 = \frac{AF}{2}\left(\sqrt{\Delta P_1 + \Delta P_2} + \sqrt{\Delta P_1 - \Delta P_2}\right) = \frac{AF}{2}\left(\sqrt{C\Delta P_1} + \sqrt{D\Delta P_1}\right) \tag{7.9}$$

It is obvious that $2 > C > 1$ and $D < 1$. $C\Delta P_1$ is used to replace $(\Delta P_1 + \Delta P_2)$. $D\Delta P_1$ is used to replace $(\Delta P_1 - \Delta P_2)$. So $(C + D)\Delta P_1 = 2\Delta P_1$. The value of $C + D$ must be equal to 2. The value of $\sqrt{C} + \sqrt{D}$ must be larger than $\sqrt{2}$ but be less than 2. Therefore, $Q > Q_3 > Q_1$. Meanwhile, Q_3 may be $<Q_2$, may also be $>Q_2$.

7.6.5 When Thermal Pressure Is Larger Than Wind Pressure

This means $\Delta P_2 > \Delta P_1$.

It is obvious that there is outflow on upper part of the window, while there is inflow on lower part of the window. It is symbolized as Q_4.

$$Q_4 = \frac{AF}{2}\sqrt{\Delta P_1 + \Delta P_2} = \frac{AF}{2}\sqrt{B\Delta P_1} \tag{7.10}$$

It is obvious that when $B > 2$, $Q_4 > Q_1$. When $B > 4$, i.e., $\Delta P_2 > 3\Delta P_1$, $Q_4 > Q_1$. Otherwise, $Q_4 < Q$.

For example, when the wind velocity is slightly larger than the static wind (which is assumed 0.35 m/s), the effect by thermal pressure with temperature difference 4.5 °C ($4.5 \times 0.0045 = 0.202$ Pa) is larger than the effect by wind pressure by three times (0.066 Pa, which can be found in the calculation later). Therefore, when there is almost no wind but only temperature difference with several degrees, $Q_4 > Q$.

Because during the haze days, there is basically static wind. When the temperature is 4.5 °C, the infiltration flow rate under the combined effect of thermal pressure and wind pressure reaches the maximum. Therefore, calculation should be performed with local maximum temperature difference between indoors and outdoors in winter. In related regulations, the outdoor design temperature for heating in winter can be obtained, such as −7.6 °C in Beijing and 17.9 °C in Sanya at maximum. The calculated outdoor temperature is recommended to be lower by 3 °C from the outdoor design temperature for heating in winter. The calculated indoor temperature in winter is assumed 20 °C. In this case, the relative large temperature difference between indoors and outdoors can be obtained. Even though in Sanya, the temperature difference between indoors and outdoors in winter will be also larger than 4.5 °C. In summer, the temperature difference will be much larger. Therefore, the infiltration rate of pollutant during the haze days in difference places can be calculated with the infiltration flow rate with the combined effect of wind pressure with velocity slightly larger than static wind (such as 0.35 m/s) and thermal pressure. Of course, it can also be calculated with the setting air velocity according to the actual situation in specific locations.

7.7 Calculated Infiltration Rate of Pollutant During Haze Episode

7.7.1 Determination of Infiltrated Particle Concentration During Haze Episode

In order to calculate the particle concentration, the infiltration flow rate should be determined at first. As mentioned before, the windward pressure should be obtained with the calculated wind velocity along the predominant wind direction in winter. In national standard, the infiltration rate for Class 4 window is determined with the pressure difference 10 Pa (the pressure difference between the windward pressure and the "0" pressure indoors). But in the haze days, the outdoor particle concentration is large while the wind velocity is small. It is usually within the static wind condition. For example, the condition from moderate haze to serious haze appeared the whole day on Nov. 29th, 2016. According to the sampling data outdoors on the 4th floor, except gust in some occasions, the wind velocity in the whole day was between 0

Table 7.7 Sampling data

Time	$PM_{2.5}$/(μg/m³)	Wind velocity/(m/s)
9:30 A.M. on Feb. 14th	120	1
15:00 P.M. on Feb. 14th	240	0.3
17:30 P.M. on Feb. 14th	223	0.3
18:20 P.M. on Feb. 14th	235	0.3
9:15 A.M. on Feb. 15th	288	0
9:15 A.M. on Feb. 16th	100.7	1.5
10:15 A.M. on Feb. 16th	18	1.8
14:30 P.M. on Feb. 16th	18	1.5
18:30 P.M. on Feb. 16th	18	1.5

and 0.3 m/s. Sometimes the wind velocity could reach more than 1 m/s. Once the wind appears, the haze will dissipate gradually.

Table 7.7 shows several sampling data outdoors on the above-mentioned 4th floor in the early of 2017.

For weather with moderate pollution or more severe, the wind velocity will be approaching 1 m/s or less, even to the static wind condition.

In Sect. 1.1, it is pointed out that the atmospheric pollution in winter is especially serious, which is related to the weather. One reason is that according to the evaluation report from Chinese Academy of Engineering cited in Chap. 1, heating is needed for the Northern China in winter. Among the 12 million tons of coal consumed in Beijing in the whole year of 2015, 75% is used for heating period. The other reason is that according to the interview of the minister at Ministry of Environmental Protection of the People's Republic of China cited in Chap. 1, the large wind frequency was smaller than 10% since 2013. The frequency of small wind and high temperature is more than 50%, and in 2016 it reached 60%. Therefore, during the calculation on the pollution concentration during haze days, this phenomenon should be paid attention to.

Suppose the wind velocity is assumed 0.35 m/s and the density ρ is 1.2 kg/m³ (of course the actual value can also be used), the windward pressure can be obtained with Eq. (7.1).

$$\Delta P_1 = 0.9\frac{(0.35)^2 \times 1.2}{2} \approx 0.066\ \text{Pa}$$

With Eq. (7.3) and the width of gap in Table 7.4, the infiltration flow rate under the wind pressure for Class 4 single window is obtained:

$$Q = 3600 \times 0.3 \times 7.5\ \text{m} \times 0.57 \times 10^{-3}\sqrt{\frac{2 \times 0.066}{1.2}} = 5.96\sqrt{0.066} = 1.53\ \text{m}^3/\text{h}$$

When the temperature difference between indoors and outdoors is assumed 4.5 °C, the thermal pressure ΔP_2 should be:

$$\Delta P_2 = 9.8 \times 4.5 \times 0.0046 = 0.202 \text{ Pa}$$

$$3\Delta P_1 = 3 \times 0.066 = 0.198 \text{ Pa}$$

$$\therefore \ \Delta P_2 > 3\Delta P_1$$

So the infiltration flow rate under the combined effect of wind pressure and thermal pressure will be larger than that with the pure wind pressure. The infiltration flow rate can be calculated with Eq. (7.8):

$$Q_4 = \frac{AF}{2}\sqrt{0.066 + 0.202} = \frac{5.96}{2}\sqrt{0.268} = 1.54 \text{ m}^3/\text{h}$$

If the temperature difference is 15 °C, the thermal pressure ΔP_2 could reach 0.68 Pa, and $Q_4 = 2.58$ m^3/h.

If the temperature difference is 30 °C, the thermal pressure ΔP_2 could reach 1.35 Pa (Since the density becomes 1.342 kg/m^3 at -10 °C, the coefficient in the expression of Q becomes 5.64), and $Q_4 = 3.36$ m^3/h.

For a bedroom with area 12 m^2 (volume 30 m^3), it takes about 20 h or even 9 h for the infiltration airflow to exchange the indoor air for one times. This means after one night, the indoor concentration will approach the outdoor concentration.

For a living room with area 20 m^2 (volume 50 m^3), when there is 1.5 Class 4 window and when the temperature difference is 30 °C, the indoor particle concentration will approach the outdoor concentration within less than 10 h.

If the Class 4 window is converted into Class 7, the width of the gap may be reduced to 0.23 mm. In this case, when the temperature difference is 4.5 °C, Q_4 will become $1.54 \times 0.23/0.57 = 0.62$ m^3/h. The infiltration flow rate for the temperature difference 30 °C is 1.42 m^3/h, which is smaller than the case with Class 4 window at temperature difference 4.5 °C. So the situation will be improved greatly.

Based on the above comprehensive analysis, when $\Delta t \geq 5$ °C, the infiltration flow rate during the haze day should be calculated with the combined effect of wind pressure and thermal pressure. The wind pressure can be calculated with the calculated wind velocity 0.35 m/s or the design value such as 1 m/s. If the value 1 m/s is used, the infiltration flow rate at temperature difference 4.5 °C will be increased by 1 m^3/h.

7.7.2 *Penetration Coefficient*

There are many expressions for the concept of penetration coefficient. For example, some symbol was used to express "the penetration coefficient of ambient particles entering into the room", which was applied to explain the extent (the ratio) of "build-

Table 7.8 Measured indoor and outdoor $PM_{2.5}$ concentrations

Time	Location	Average $PM_{2.5}$ concentration ($\mu g/m^3$)	I/O ratio
2014.2.24	Outdoor (Third Ring Road of Beijing)	772	0.72
	Indoor (with window closed)	556	
2014.3.3	Outdoor (Third Ring Road of Beijing)	529.7	0.73
	Indoor (with window closed)	388.3	

ing on penetration coefficient of particles". Or it was used to illustrate "the ratio of the particle mass concentration entering into the room with the infiltration airflow through the building envelop". It is impossible for the infiltration airflow to penetrate the physical enclosure structure. It should be the gap, or mainly the gap of door and window. This must appear in the definition of terms. Even though it is explained as the gap later, it is not rigorous. Moreover, it is not only the wind that causes the infiltration. If there is ventilable gap, it is incredible (except the temporary building).

Therefore, it is especially pointed out that "penetration coefficient means the ratio between the particle concentration infiltrated indoors through gaps of door and window and the outdoor concentration". So it is more appropriate to use the term "penetration coefficient of gap". The original term will be kept since here in the book.

During the haze days, particles will not penetrate through the gaps of windows completely and then enter into the room. Some of them will deposit inside. Since the structures of window gap are different, the real situation cannot be reflected no matter in experiment and simulation (sometimes only rectangular straight gap is used).

For safety purpose, one opinion is to perform calculation with full penetration. The other is to provide a penetration coefficient.

Table 7.8 shows the measured result by author's affiliation. The so-called penetration coefficient can be considered to 0.75. This means the final indoor particle concentration can reach 75% of outdoor value. In Chinese standard GB/T18801-2015 "*Air cleaner*", the penetration coefficient is assumed 0.8. Both are fine. For safety reason, it seems feasible to use full penetration.

The general formula of the infiltration flow rate under the combined effect of wind pressure and thermal pressure during the haze day can be expressed as follows:

When the temperature difference is smaller than 15 °C:

$$Q_4 = 0.7 \times \text{Gap length (m)} \times \text{The maximum width of gap (mm)} \times \sqrt{0.066 + \Delta P_2} \quad \text{m}^3/\text{h} \tag{7.11}$$

When the temperature difference is larger than 15 °C:

$$Q_4 = 0.66 \times \text{Gap length (m)} \times \text{The maximum width of gap (mm)} \times \sqrt{0.066 + \Delta P_2} \quad \text{m}^3/\text{h} \tag{7.12}$$

For different places, the value of ΔP_2 is changed. The variation of the coefficient 0.066 is very small, which can be assumed constant.

References

1. H. Zhao, H. Zhai, Determination of calcualted wind velocity for infiltration in China with safety probability method. J. HV&AC **1**, 16–20 (1994)
2. Z. Xu, *Fundamentals of Air Cleaning Technology*, 4th edn. (Science Press, Beijing, 2014)
3. GB50736-2012, *Design Code for Heating Ventilation and Air Conditioning of Civil Buildings*. (China, 2012)
4. Z. Xu, *Design Principle of Isolation Ward* (Science Press, Beijing, 2006)
5. Japan Air Cleaning Association, *Air Cleaning Handbook* (1981)

Chapter 8
Calculation on Cleaning Capability of Air Purifier [1]

8.1 Outline

8.1.1 Necessity of Calculation

After application of air purifier, what kind of air pollution level inside the room can be maintained with the corresponding outdoor pollution condition? This is the necessary condition to evaluate the capability of air purifier. It is also one of the concerns for users. It is one of the main problems the designer and the manufacturer have to answer clearly. The calculation method for the capability of air purifier is also applicable to the calculation of air cleaning system.

8.1.2 Difficulty During Calculation

The most difficult problem during the calculation on the air purification capability of air purifier is that the efficiency for $PM_{2.5}$ is unknown. The second problem is whether the concept of "non-guarantee rate" should be adopted. If yes, how to determine the value of the non-guarantee rate? For other problems, the clue can be found in the principle of air cleaning technology.

8.2 Determination of Calculation Parameters

According to Eq. (5.16), parameters related to the calculation on indoor particle concentration include:

Z. Xu, *Air Purifier*, https://doi.org/10.1007/978-981-13-2532-8_8

η is the efficiency of air filter;
M is the concentration of atmospheric dust;
G is the particle generation rate indoors per unit volume;
n is the air change rate;
S is the return air ratio;
ψ is the non-uniform distribution coefficient.

They will be determined separately in the following sections.

8.2.1 About η

During the discussion of air purifier and haze, it is inevitable to mention $PM_{2.5}$. How could we sample and obtain the gravimetric concentration of particles with aerodynamic diameters ≤2.5 μm? Because optical particle counter is based on the particle counting method for particles with size equal to or larger than certain value, people is always confused by the problem of $PM_{2.5}$ efficiency during the calculation and analysis of air purifier and haze. Now through qualitative analysis, three alternative methods will be discussed. The first one is to replace the $PM_{2.5}$ particle counting efficiency by the particle counting efficiency with the commonly used size ≥0.5 μm. The second one is to replace the $PM_{2.5}$ particle counting efficiency by the most penetrating particle size (MPPS) efficiency. The third one is to gravimetric efficiency of $PM_{2.5}$ by the particle counting efficiency with the size ≥0.5 μm.

8.2.1.1 Replacement with the Particle Counting Efficiency with the Size ≥0.5 μm

The efficiency of air filter commonly used in air purifier can be expressed by the particle counting efficiency with the size ≥0.5 μm. The sodium flame efficiency of type A HEPA filter in China should be ≥99.9% at least. It is equivalent with the particle counting efficiency more than 99.99% for the size ≥0.5 μm. The sodium flame efficiency of type B HEPA filter in China should be ≥99.99%. It is equivalent with the particle counting efficiency more than 99.999% for the size ≥0.5 μm.

There is a misunderstanding that since the air filter has certain particle counting efficiency for the size ≥0.5 μm, there is no efficiency for particles with size less than 0.5 μm. It is shown for the Chinese filter media in Fig. 8.1 and U.S. filter media in Fig. 8.2 [2] that the efficiency for particle size 0.1 μm is still close to that for 0.5 μm.

Table 8.1 shows the efficiency of polypropylene fibrous media [2]. Efficiency for particle size 0.4 μm is larger than that for particle size less than 0.4 μm. The minimum efficiency is close to the particle size 0.05 μm. So efficiency for particle size larger than 0.5 μm is much larger. Therefore, it is safe and feasible to design, calculate and compare air purifiers based on the particle counting efficiency with size ≥0.5 μm under normal condition.

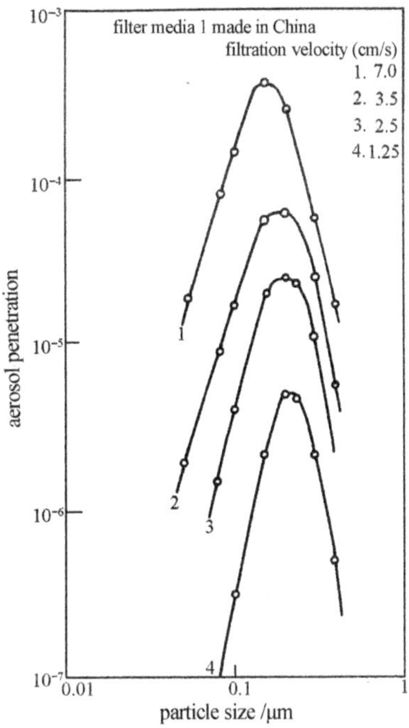

Fig. 8.1 Relationship between particle size and penetration for filter media 1 (common HEPA filter) made in China

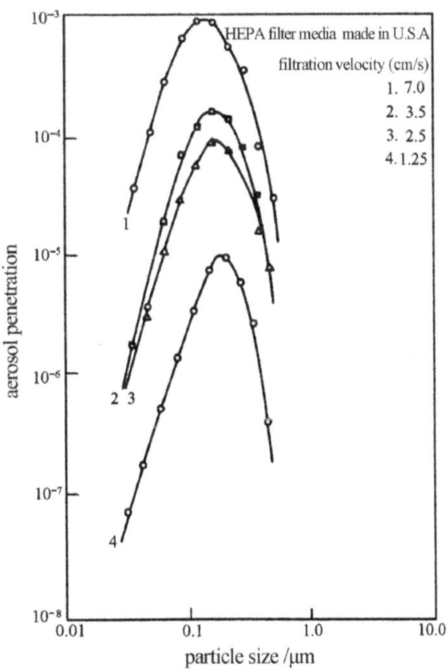

Fig. 8.2 Relationship between particle size and penetration for filter media (common HEPA filter) made in U.S.A

Table 8.1 Efficiency with DOP for domestic made polypropylene fibrous media

ΔP/Pa	20.3	40.6	59.9	116.8	482.6
v/(cm/s)	1.25	2.5	3.5	7.0	13.4
Particle size/μm	Efficiency/%				
0.5	–	–	–	–	99.83
0.4	99.9994	99.995	99.984	99.83	99.59
0.3	99.9991	99.990	99.970	99.71	99.12
0.2	99.9950	99.970	99.920	99.41	98.37
0.15	99.9600	99.870	99.780	98.48	97.78
0.10	99.8800	99.660	99.370	98.62	96.26
0.08	99.8500	99.560	99.270	96.52	95.39
0.05	99.8400	99.400	99.170	95.54	93.80
0.03	99.9700	99.850	99.710	98.52	–

8.2.1.2 Replace of $PM_{2.5}$ Particle Counting Efficiency with MPPS Efficiency

The MPPS efficiency is the minimum efficiency of HEPA and ULPA filters. This value is much lower than the efficiency with much smaller particle size. It is shown in Figs. 8.1 and 8.2 that the MPPS size of HEPA filter is usually between 0.2 and 0.3 μm. The theoretical MPPS size is also between 0.2 and 0.3 μm [2]. The MPPS efficiency is ≥99.9%. Therefore, when the MPPS efficiency of HEPA filter is known, it is safe to calculated $PM_{2.5}$ efficiency when this efficiency value is used to represent the particle counting efficiency of particles with size ≤2.5 μm. As for ULPA filter, the MPPS size is much smaller, which is about 0.15–0.2 μm. One example for the standard from China Refrigeration and Air-conditioning Industry Association CRAA431.1-2008 is shown in Fig. 8.3.

8.2.1.3 Replacement of Gravimetric Efficiency of $PM_{2.5}$ by the Particle Counting Efficiency with the Size ≥0.5 μm

During the discussion of haze, if the gravimetric efficiency for particles with size ≤2.5 μm can be used, it may be more intuitive for ordinary user, but it will usually not influence the judgment of professionals. When the distribution of atmospheric dust during the haze days is unknown and the field measurement data is lacking, the following qualitative analysis will be given.

Since the common distribution of atmospheric dust is different from that during the photo-chemical reaction pollution such as haze days, there is lack of comparability when the $PM_{2.5}$ efficiency is obtained by experiment with pure aerosol.

The distribution of atmospheric dust under normal condition is shown in Table 8.2 [3]. The proportion for the mass of particles with diameter less than 0.5 μm is only

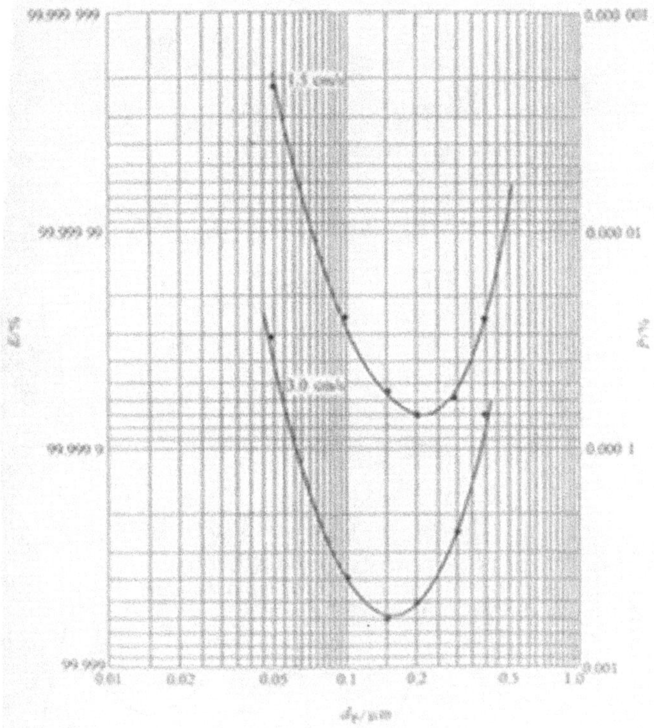

Fig. 8.3 Relationship between efficiency E, penetration P and particle diameter d_p with two air velocities for ULPA filter media (example)

1% of the total mass. The unimodal mass distribution appears between 5 and 10 μm. But during the photo-chemical smog pollution, it is bimodal distributed, which is shown in Fig. 8.4 [3].

Usually the arrestance of air filter will be larger than the particle counting efficiency. For example, experiment shows that when the particle counting efficiency for particle size ≥0.5 μm reaches more than 80%, the arrestance may approach 97%. This is because the filtration efficiency of air filter for micron particles is much larger than that for sub-micron particles. For HEPA filter, the efficiency is almost close to 100%. For air filters with particle counting efficiency more than 99.9%, it is no doubt that the arrestance value can be considered as equivalent with the particle counting efficiency. This is the main basis for the following discussion.

It is shown from Fig. 8.4 that the proportion for the mass of particles with size more than 0.5 μm is about more than 2/3 of the whole size range. As mentioned before, the particle counting efficiency with size ≥0.5 μm is used as the basis, which dose not mean that there is no efficiency for particle size <0.5 μm. Except for the particle size range 0.2–0.25 μm, the efficiency may be lower by one order of magnitude. The efficiency for smaller particle size is comparable to that with particle size 0.5 μm.

Table 8.2 Distribution for number and mass of atmospheric dust

Particle size interval/μm	Average size/μm	Number/%		Mass/%	
		Percentage	More than 0.5 μm are considered as 100	Percentage	More than 0.5 μm are considered as 100
0–0.5	0.25	91.68	–	1	–
0.5–1	0.75	6.78	81.49	2	2.02
1–3	2	1.07	12.86	6	6.06
3–5	4	0.25	3	11	11.11
5–10	7.4	0.17	2	52	52.53
10–30	20	0.05	0.65	28	28.28

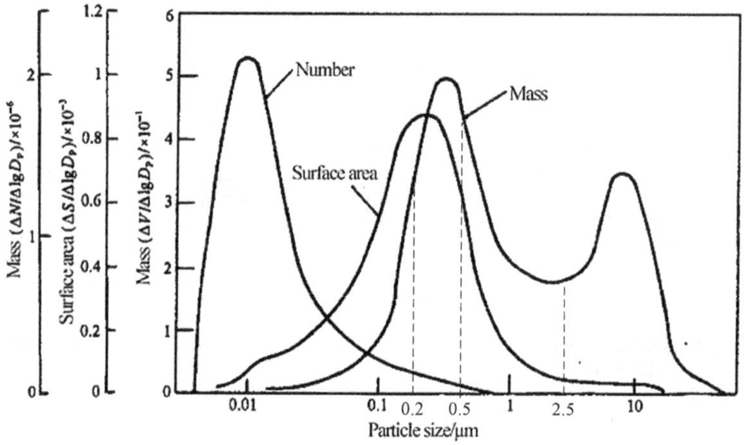

Fig. 8.4 Example of full particle size distribution of atmospheric dust during photo-chemical smog pollution (the dashed lines are added by author)

Therefore, the arrestance of HEPA filter can be supposed to be the particle counting efficiency with size ≥0.5 μm. The difference is not very large. But difference will be very large for non-HEPA filter.

It is shown in Fig. 8.4 that the ratio for the mass of the particles with particle size smaller than MPPS size 0.2 μm is about less than 10%. If the minimum particle counting efficiency for size ≥0.2 μm and the comparable arrestance are used for calculation, there will be larger safety factor.

The above is related to the whole particle size distribution.

As shown in Fig. 8.4, if the particle counting efficiency with size ≥2.5 μm is used to represent the arrestance with $PM_{2.5}$, i.e., particle size ≤2.5 μm, the value is obviously larger. But if the particle counting efficiency with size ≥0.2 μm is used to represent the arrestance with $PM_{2.5}$, it is obviously smaller (In this case, of course it is safe).

From the figure, the gravimetric concentrations for the particle size ranges 0.01–0.5 μm, 0.5–2.5 μm, and ≥2.5 μm are roughly equivalent. The particle counting efficiency with size ≥0.5 μm is larger than the particle counting efficiency with size ≤2.5 μm (because there is the minimum efficiency existing in this range). The particle counting efficiency with size ≤2.5 μm is smaller than the arrestance with size ≤2.5 μm. So the particle counting efficiency with size ≥0.5 μm may be comparable to the arrestance with size ≤2.5 μm. This situation may occur for filters with performance better than sub-HEPA filter, especially HEPA filter during the haze episode. It is shown in Table 8.2 that for the distribution of atmospheric dust during the normal day, the proportion for the mass of particles with size ≤2.5 μm only occupies about 9% of the total mass. The proportion for the mass of particles with size ≥0.5 μm occupies 99% of the total mass. The above conclusion cannot be obtained in this case.

For fine filter (the particle counting efficiency with size ≥0.5 μm is about 50%) during the haze days, the particle number for size >2.5 μm will be much smaller than that for size <0.5 μm. So the particle counting efficiency for size ≥0.5 μm may be less than that for size ≤2.5 μm. The lower the efficiency of this air filter is, the more obvious this phenomenon will be. The resultant particle counting efficiency with size ≥0.5 μm may also be less than the arrestance with size ≤2.5 μm.

The analysis based on Fig. 8.4 is only one example during the actual pollution day. The analysis may be different from the particle size distribution in other pollution day.

Figure 8.5 shows one example measured by Tu et al. [4]. The liquid aerosol was used as artificial aerosol to measure the particle counting efficiency with size ≥0.5 μm. The test air was used to measure the arrestance of $PM_{2.5}$. It is shown that for air filters with efficiency more than 70% with particle size ≥0.5 μm (belong to high and medium efficiency air filter), the arrestance with size ≤2.5 μm is comparable to the particle counting efficiency with size ≥0.5 μm. For air filters with the particle counting efficiency <70% with size ≥0.5 μm, the above arrestance will be higher than the above particle counting efficiency. The smaller the efficiency of air filter is, the more obvious the phenomenon will be. For example, when the particle counting efficiency for size ≥0.5 μm is 50%, the arrestance for size ≤2.5 μm is 56%. When the particle counting efficiency for size ≥0.5 μm is 20%, the arrestance for size ≤2.5 μm is 34%. When the particle counting efficiency for size ≥0.5 μm is 10%, the arrestance for size ≤2.5 μm is doubled to be 26%.

This test case is basically consistent with the above qualitative analysis.

Although it was not mentioned whether it was haze day during experiment, it is the slight mild haze day since the average $PM_{2.5}$ concentration 116.3 $\mu g/m^3$ can be obtained from the given concentration range 88–159 $\mu g/m^3$. It means that the example shown in Fig. 8.4 is equivalent to the mild to moderate haze pollution. If measurement can be performed during severe haze pollution days, it will be helpful for explaining the problem.

Since the particle size distributions in haze days are also different, in situ measurement data during different haze days should be provided. By analysis, the data and its correction method for most situations are likely to be obtained and adopted. From

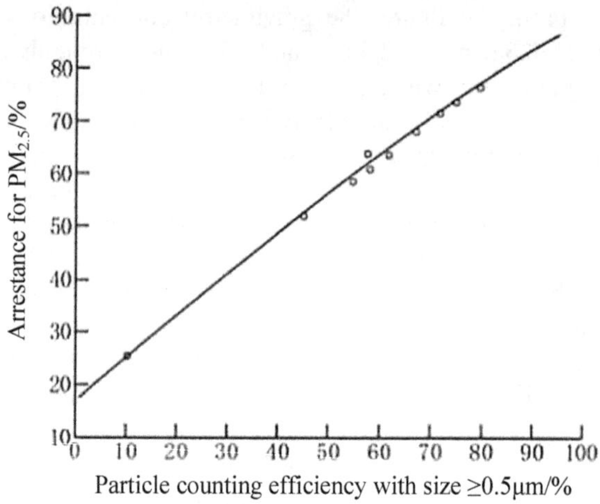

Fig. 8.5 Relationship between the particle counting efficiency with size ≥0.5 μm and the arrestance for size ≤2.5 μm for air filter

the above-mentioned qualitative analysis, if the proportion for the mass of particles with size less than 2.5 μm is larger, the particle counting efficiency for size ≥0.5 μm with high efficiency filter will be less than the arrestance with $PM_{2.5}$. The difference may be larger between the particle counting efficiency and the arrestance for low efficiency filter.

Since air filters with performance better than sub-HEPA filter is usually used in air purifier, the particle counting efficiency with size ≥0.5 μm can be used to calculate and evaluate the purification performance of air purifier in terms of arrestance during haze days.

The measurement result in Fig. 8.5 is almost a straight line. When straight line is used, the empirical equation fitted by author is:

$$y = 17.5\% + 0.76 \cdot x \tag{8.1}$$

where y is the arrestance of $PM_{2.5}$, %;
x is the particle counting efficiency of particles with diameter ≥0.5 μm, %;
17.5% is the intercept in the figure.

Example 1 When the particle counting efficiency of particles with diameter ≥0.5 μm is 45.2%, what is the value of y?

Ans: $y = 17.5\% + 0.76 \times 45.2\% = 51.9\%$.
The measured value is 52.3%.

Example 2 When the particle counting efficiency of particles with diameter ≥0.5 μm is 75%, what is the value of y?

Ans: $y = 17.5\% + 0.76 \times 75\% = 74.5\%$.
The measured value is 74.5%.

8.2.2 *About* M

8.2.2.1 Particle Counting Concentration

According to analysis in Chap. 5, when the particle counting concentration with the size ≥0.5 μm reaches 10^6 pc/L, the standard for pollution is reached. However, it is required in national standard "*Air Cleaner*" that the original particle counting concentration with the size ≥0.3 μm during test should reach 10^6–10^7 pc/L. So it is recommended here to adopt the particle counting concentration with the size ≥0.5 μm reaches 10^7 pc/L. Of course, this is only a suggestion.

8.2.2.2 Gravimetric Concentration

When the gravimetric concentration of $PM_{2.5}$ is used for calculation, how to determine the outdoor $PM_{2.5}$ concentration? Is calculation performed based on the non-guarantee rate or non-guarantee day? In author's opinion, it has no implication for the application of air purifier.

(a) It is only recent four years for the monitoring of $PM_{2.5}$ concentration.
(b) The characteristic of $PM_{2.5}$ during haze day is not so obvious as the distribution of atmospheric dust concentration during normal weather. It is influenced by weather and pollution source. The contingency rate is very large. The stability is much poorer than that of temperature and humidity.
(c) The value of outdoor $PM_{2.5}$ concentration will only influence the self-purification time after the operation of air purifier and when the standard stable concentration is reached. For air purifier containing the air filter with performance better than sub-HEPA filter, even through the outdoor concentration is extremely large, it is also very easy for the indoor concentration to reach the standard concentration value. Meanwhile, it will not consume more cost and energy. For the general air conditioning system with air change rate 10 h^{-1}, when air filters with performance better than high-medium efficiency filter is used and when M is not larger than 365 μg/m^3, the indoor concentration can meet the requirement of standard. This can be seen from the following calculation. The value of M has no substantial impact on the equipment.
(d) When calculation is performed on the basis of non-guarantee rate, such as 5% of the non-guarantee rate, the design concentration obtained for severe polluted cities with haze such as Beijing may only be two hundred and dozens μg/m^3. But according to the report from newspaper on Feb. 26th in 2014, there were 58 cities among 161 monitored cities in China, where the concentration was larger than 150 μg/m^3. The concentration in 23 cities was larger than 250 μg/m^3. Average pollution concentration in Beijing even reached 333 μg/m^3. Sometimes the single measurement in some places may even larger than this value, which reached more than 500 μg/m^3. In the same year, the measured value in the yard near the North 3rd Ring Road of Beijing on Feb. 24th even reached 772 μg/m^3. Between

Dec. 16th and Dec 21st of 2016 in Beijing, the measured values always “beyond the test range of the instrument” (which means it reached 500 $\mu g/m^3$). In some place, it even reached 1000 $\mu g/m^3$. According to “2016 China Environmental Status Bulletin” issued by Ministry of Environmental Protection of the People’s Republic of China, for 338 perfecture-level and above cities in China in 2016, there were 254 cities where the concentration in the ambient air was larger than the standard value. It occupies 75.1% of these cities.

Equipment such air purifier is mainly used to deal with the polluted weather. As the evaluation report by Chinese Academy of Engineering in Chap. 1, there are 23.3%, i.e. 85 days, among 338 cities in statistics when the average concentration was larger than the standard value. According to “2016 China Environmental Status Bulletin”, this ratio in 2016 was reduced to 21.2%, i.e., 74 days. Since the air purifier will not be used in the whole year, it is safe to consider that it is used in serious polluted weather (the situation is worse than severe polluted weather) during the total usage of less than 100 days. As mentioned in Chap. 1, according to Chinese standard, it is severe pollution when the 24 h average concentration of $PM_{2.5}$ exceeds 150 $\mu g/m^3$. It is serious pollution when the corresponding concentration exceeds 250 $\mu g/m^3$. For safety reason, the above average concentration in Beijing is 333 $\mu g/m^3$. So the value of 350 $\mu g/m^3$ can be adopted for Beijing. Or even the value 500 $\mu g/m^3$ can be used.

Air purifier will not be classified with “non-guarantee rate”. Users will also not operate the air purifier with “non-guarantee rate”. This is different from the cooling and heating loads. Therefore, it is suitable to set the parameter M with the above value.

8.2.3 About G

(1) When outdoor air in introduced into the air purifier, there is usually a certain value of positive pressure in the room. In this case, the infiltration through the gap cannot be considered. There is only particle generation source indoors.

It is quite complex for the particle generation indoors. But one simple and safe situation can also be considered. Correction can be made in the practice in future.

For occupant with ordinary cloth, the average particle generation rate with various activities is 2.14×10^5 pc/(min·person) [5]. It is more difficult to determine the particle generation rate from various surfaces. Here only assumption can be made at first.

The occupant density is assumed 0.15 person/m^2. For the small living room with area 20 m^2, there is three people. According to the calculation for cleanroom, the particle generation rate from 8 m^2 clean floor is equivalent to one people with clean cloth. So we assume that the particle generation rate from 8 m^2 non-clean floor is equivalent to one people with ordinary cloth. Then the room with area 20 m^2 can be consider to contain six people. The corresponding occupant density becomes 0.3. This means that the occupant density is doubled.

When there is only one people per m^2, the generation rate per unit volume (room height 2.5 m) becomes:

$$\frac{2.14 \times 10^5}{2.5 \times 1} = 0.86 \times 10^5 \text{ pc/(m}^3\cdot\text{min)}$$

When the particle generation from the surface is considered, the equivalent occupant density can be considered as 0.15×2. In this case, the generation rate per unit volume becomes

$$G = 0.15 \times 2 \times 0.86 \times 10^5 = 0.26 \times 10^5 \text{ pc/(m}^3\cdot\text{min)}$$

During calculation process, various temporary particle generation source should not be considered. For example during the cooking process, oil fume can be exhausted by the range hood temporarily. Otherwise, it will be more difficult if the abrupt increase of particle is considered, which has no practical implication.

From the value of G, it is much smaller than the concentration of outdoor air. Moreover, during the normal condition, there is no indoor pollution so high as the haze days. So it can also be omitted (such as the national standard on air purifier).

(2) Indoor air is returned with self-circulation. G can be classified as G_1 and G_2.

   (a) Suppose the indoor particle generation rate is, it is equal to the above value: $G_1 = 0.26 \times 10^5$ pc/(m^3·min)
   (b) The infiltrated particles from outside can also be considered as the particles generated from indoors. The corresponding generation rate is symbolized as G2.

According to the previous analysis, for a living room with area 20 m^2 and height 2.5 m, there is a window with performance equivalent to Class 4. Under the combined effect of thermal pressure (temperature difference 30 °C) and wind pressure (wind velocity 0.35 m/s), the infiltration flow rate of polluted air for each Class 4 window is 3.36 m^3/h according to Sect. 7.8. The infiltrated particle counting concentration should be:

$$G_2 = \frac{p \times q \times M}{60 \times F \times H} \tag{8.2}$$

where p is the penetration coefficient, which can be set 0.75;
q is the infiltration flow rate with the combined effect of wind and thermal pressures (based on temperature difference 30 °C, which is shown in the first item in Sect. 7.8), m^3/h;
M is the particle counting concentration of atmospheric dust outdoors, pc/L;
F is the room area, m^2;
H is the room height, m.

When values are inserted, we can obtain:

$$G_2 = \frac{0.75 \times 3.36 \times 10^7 \times 10^3}{60 \times 20 \times 2.5} = 0.84 \times 10^7\ \text{pc}/\left(\text{m}^3 \cdot \text{min}\right)$$

As mentioned before, when penetration coefficient 0.75 can be considered for the above equation, or the complete penetration can be considered.

The gravimetric concentration becomes:

$$G_2 = \frac{0.75 \times 3.36 \times 350}{60 \times 20 \times 2.5} = 0.3\ \mu\text{g}/\left(\text{m}^3 \cdot \text{min}\right)$$

8.2.4 *About* η

The volume of the living room is assumed $20 \times 2.5 = 50$ m^3.

When the nominal flow rate of air purifier is 300 m^3/h, the air change rate is $n =$ 6 h^{-1}. When the nominal flow rate of air purifier is 400 m^3/h, the air change rate is $n = 8$ h^{-1}.

8.2.5 *About* S

The value of S is determined by the fresh air ratio $(1 - S)$.

$$1 - S = \frac{Q}{Q_0} \tag{8.3}$$

where Q is the flow rate of outdoor air, m^3/h. For ordinary condition, it is based on the requirement of dilution CO_2 in residential house and public space;
Q_0 is the total flow rate of supply air, m^3/h.

$$Q = \frac{L}{(C - C_0) \times 10^{-3}} \tag{8.4}$$

where L is the CO_2 generation rate in the room, m^3/h. It is shown in Table 8.3[6].

C_0 is the CO_2 concentration in the atmosphere, L/m^3; The measured value in city is usually larger than 0.3 L/m^3. It can be set 0.4 L/m^3;

C is the controlled CO_2 concentration in the room; According to Chinese national standard GB/T18883 "*Indoor Air Quality*", it is required to be 0.10%, which is 1 L/m^3; When the quite status is considered, the exhaled CO_2 volume is set 0.013 m^3/(p·h) according to Table 8.3.

So we obtain:

Table 8.3 Relationship between the labor intensity and exhaled CO_2 volume for male

Labor intensity	Exhaled CO_2 volume/[m^3/(p·h)]	Adopted exhaled CO_2 volume during calculation/[m^3/(p·h)]
Quite	0.0132	0.013
Extremely mild labor	0.0132–0.0242	0.022
Mile labor	0.0242–0.0352	0.03
Moderate labor	0.0352–0.0572	0.046
Heavy labor	0.0572–0.0902	0.074

$$Q = \frac{0.013}{(1-0.4)\times 10^{-3}} = \frac{0.013}{0.0006} = 22\ \text{m}^3/\text{h}$$

If there are three people, $Q = 66$ m^3/h.

So if $Q_0 = 300$ m^3/h, $1 - S = 0.22$. If $Q_0 = 400$ m^3/h, $1 - S = 0.165$.

This ratio of outdoor air can meet occupant's requirement. Moreover, according to Sect. 7.7, if there is 1.5 Class 4 external windows in the living room containing three people, the infiltration flow rate during heavy haze days (which means the wind velocity is extremely small) under the temperature difference 30 °C becomes $1.5 \times 3.36 = 5$ m^3/h. If the pressure difference 10 Pa in Table 7.3 is used for calculation, the maximum infiltration flow rate is only 28 m^3/h. These two infiltrated airflow can be counteracted by the positive pressure formed by the flow rate of outdoor air, which makes them impossible to enter into the room.

The flow rate of outdoor air mentioned above is far beyond the requirement of GB50736-2012 "*Design Code for Heating Ventilation and Air Conditioning of Civil Buildings*". Table 8.4 shows the requirement of this standard.

The area of this living room is 20 m^2, and the volume is 50 m^3. According to Table 8.4, the air change rate of outdoor air is 0.5 h^{-1}, which is 25 m^3/h. If there are three people living in the apartment (two bedrooms and one living room), the total area is $20 + 15 + 15 = 50$ m^2. The volume is 125 m^3. When the applicable air change rate of outdoor air is still 0.5 h^{-1}, the flow rate becomes 62.5 m^3/h. If two doors of two bedrooms are fully open, the above flow rate 66 m^3/h of outdoor air also meets the requirement. It is shown from Fig. 8.6 that the minimum air change rate of outdoor air in two thirds of different national standards is 0.5 h^{-1}, which is counted by Zhang et al. [6]

Table 8.4 Minimum air change rate of outdoor air in residential building

Living area per person F_p	Air change rate per hour, h^{-1}
$F_p \le 10$ m^2	0.70
10 m^2 $< F_p \le 20$ m^2	0.60
20 m^2 $< F_p \le 50$ m^2	0.50
$F_p > 50$ m^2	0.45

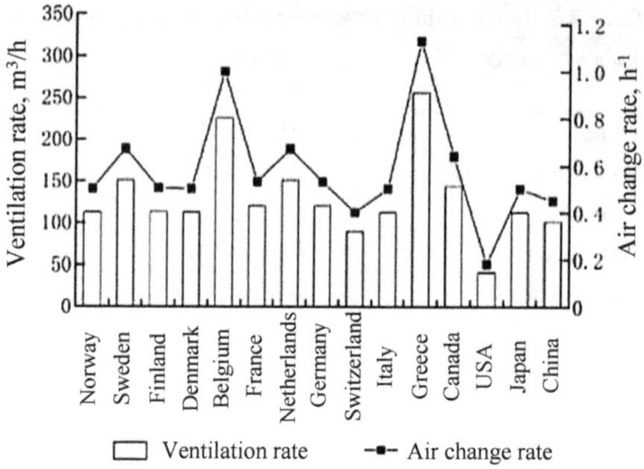

Fig. 8.6 Minimum flow rate and air change rate for residential buildings in different countries

8.2.6 *About* ψ

According to Chap. 5, when n = 6 and 8 h^{-1}, ψ can be set 1.9 and 1.7, respectively.

8.3 Calculation on *N* with Outdoor Air Condition

8.3.1 *Particle Counting Concentration*

$$N = \psi \frac{60G \times 10^{-3} + Mn(1 - S)(1 - \eta_n)}{n[1 - S(1 - \eta_r)]} \tag{8.5}$$

When $n = 6\ h^{-1}$, ψ can be set 1.9 (temporarily based on cleanroom, which is the same later).

Since the indoor pressure is positive, the infiltration by outdoor air will not be considered. And only indoor particle generation source is considered.

From (1), it is known that $G = G_1 = 0.26 \times 10^5$ pc/(m^3·min). Since $1 - S = 0.22$, $S = 0.78$. The particle counting efficiency of air filter for outdoor air is equal to that for return air, i.e., $\eta_n = \eta_r = 0.9$.

Since the indoor pressure is positive, the infiltrated airflow will not be considered. So there is no problem related to the so-called penetration coefficient.

$$N = 1.9 \frac{60 \times 0.26 \times 10^5 \times 10^{-3} + 10^7 \times 6 \times (1 - 0.78)(1 - 0.9)}{6[1 - 0.78(1 - 0.9)]}$$
$$= 1.9 \frac{15.6 \times 10^2 + 0.132 \times 10^7}{6 \times 0.922} = 1.9 \frac{0.0156 \times 10^5 + 13.2 \times 10^5}{5.53}$$
$$= 4.59 \times 10^5\ \text{pc/L}$$

$\eta_n = \eta_r = 0.99$ (particle counting efficiency)
So we obtain

$$N = 1.9\frac{0.0156 \times 10^5 + 1.32 \times 10^5}{6 \times 0.9922} = 1.9\frac{1.336 \times 10^5}{5.953} = 0.43 \times 10^5 \text{ pc/L}$$

For the air purifier with efficiency $\eta = 0.9$, the indoor particle concentration can be reduced to 1/20 of the ambient concentration. So it is also applicable for house-hold application. For air purifier with efficiency $\eta = 0.99$, the indoor particle concentration can be further reduced by one order to magnitude. This is of course much better.

If M is 10^6, N can also be almost reduced by 90%.

8.3.2 Particle Mass Concentration

If the indoor particle generation rate G_1 in terms of gravimetric concentration is much smaller than M and it is difficult to obtain the converted value in different unit, it can be omitted. Then we obtain:

$$N = \psi\frac{\mathrm{Mn}(1 - \mathrm{S})(1 - \eta_n)}{\mathrm{n}[1 - \mathrm{S}(1 - \eta_r)]} = 1.9\frac{350 \times 6 \times (1 - 0.78)(1 - 0.9)}{6[1 - 0.78(1 - 0.9)]} = 16\,\mu\text{g/m}^3$$

If high-medium efficiency air filter is used, the minimum efficiency can be assumed 0.7. $N = 57.2\ \mu\text{g/m}^3$, which is equivalent to the level of "good" for ambient air. The application will not be influenced. So for general air conditioning system, the total efficiency of air filter should not be lower than high-medium efficiency air filter.

From the above-mentioned calculation, when there is introduction of outdoor air and no infiltration airflow, the indoor particle concentration in terms of gravimetric concentration will vary between 10 and 30 $\mu\text{g/m}^3$ for the variation range of ambient concentration 200–700 $\mu\text{g/m}^3$. There is no practical implication for the so-called non-guarantee rate.

8.3.3 Calculation on Self-cleaning Time

Based on the above calculated steady-state concentration, the self-cleaning time will be calculated here. From the calculation graph of self-cleaning time in Fig. 6.1, N_0 in the expression N_0/N is the initial concentration indoors. The calculation result with $N_0 = 0.75$ M is shown in Table 8.5.

It is shown in Table 8.5 that with the same condition of n and η, the self-cleaning times needed based on the calculation with particle counting efficiency and gravimetric concentration are equivalent (please see No. 1 and No. 4). This also means that the particle counting concentration 10^7 pc/L is equivalent with the gravimetric concentration 350 $\mu g/m^3$. If the indoor particle concentration is not needed to be as low as 16 $\mu g/m^3$, the time needed of course will be much shorter.

It is shown in the above table that it is more beneficial to shorten the self-cleaning time with the increase of air change rate n than that with the increase of η. For the situation with introduction of outdoor air, when η is increased by 1.1 times, N can be reduced by 10 times. Therefore, for the situation with introduction of outdoor air, the effect of the increased η on reduction of indoor particle concentration will be much larger than the effect of the increased n. This is consistent with the above-mentioned analysis. The above calculation is considered based on non-uniform distribution in the whole room. If the main area is focused on where the supplied air will arrive at, the self-cleaning time in this area will be much smaller than the value in the table.

8.4 Calculation on *N* with Indoor Self-circulation

8.4.1 *Particle Counting Concentration (Non-fully Penetration)*

Since there is no positive pressure, outdoor pollutant will infiltrate into the room through the window gap. The infiltrated particles can be considered as the indoor particle generation source with the rate G_2.

$$N = \psi \frac{60(G_1 + G_2) \times 10^{-3}}{n\eta} \quad (8.6)$$

For the same condition as the above example, when the value of G_2 under the condition of non-fully penetration shown in Sect. 8.2 is adopted, we can obtain:

Table 8.5 Calculation result on self-cleaning time in terms of particle counting efficiency

No.	n/h^{-1}	η/% (particle counting efficiency)	N_0/N	$n\eta t$ in Fig. 6.1	$t = \frac{n\eta t}{n\eta}$/min
1	6	0.9	$\frac{0.75\times10^7}{4.59\times10^5} = 16.34$	≈450	$\frac{450}{6\times0.9} = 83$
2	8	0.9	$\frac{0.75\times10^7}{2.39\times10^5} = 31.38$	≈470	$\frac{490}{8\times0.9} = 68$
3	6	0.99	$\frac{0.75\times10^7}{0.43\times10^5} = 174.4$	≈585	$\frac{585}{8\times0.99} = 74$
4	6	0.9	$\frac{0.75350}{16} = 16.4$	≈450	$\frac{450}{5.4} = 83$

$$N = 1.9\frac{60(0.26 \times 10^5 + 0.84 \times 10^7) \times 10^{-3}}{6 \times 0.9} = 1.9\frac{60 \times 84.26 \times 10^5 \times 10^{-3}}{5.4}$$
$$= 1.9 \times 0.94 \times 10^5 = 1.8 \times 10^5\ \text{pc/L}$$

Based on the specific calculation result, the self-cleaning time can be obtained, which is shown in Table 8.6.

It is shown in the above table that for the situation of no introduction of outdoor air and self-circulation of indoor air, it is more beneficial to shorten the self-cleaning time and reduce the indoor particle concentration with the increase of air change rate n than that with the increase of η. Moreover, steady-state indoor particle concentration with much lower value can be obtained. But the self-cleaning time to reach this concentration value will be slightly larger.

8.4.2 Particle Mass Concentration for $PM_{2.5}$

Based on the above calculation, G_1 can be omitted compared with G_2 (such as the national standard on air purifier). So there is no difficulty to convert the value of the gravimetric concentration. According to Sect. 8.2.3, when the ambient concentration is 350 μg/m^3 and the penetration coefficient is 0.75, G_2 =0.3 μg/(m^3·min). So for the condition that n = 6 h^{-1} and the arrestance η=0.9, we obtain:

$$N = \psi\frac{60G_2}{n\eta} = 1.9\frac{60 \times 0.3}{6 \times 0.9} = 6.3\ \mu\text{g/m}^3$$

If M is assumed 500 μg/m^3, N approaches 10 μg/m^3.

Since n = 8 h^{-1}, N=4.8 μg/m^3.

For the calculation expression of particle counting concentration in above Eq. (8.6), there is the item 10^{-3}. This is caused by the conversion to pc/L. Now the unit is pc/m^3, so there is no 10^{-3}.

Table 8.6 Calculation result on self-cleaning time based on particle counting concentration

No.	n/h^{-1}	η/% (particle counting efficiency)	N_0/N	$n\eta t$ in Fig. 6.1	$t = \frac{n\eta t}{n\eta}$/min
1	6	0.9	$\frac{0.75\times10^7}{1.8\times10^5} = 41.7$	≈500	$\frac{500}{6\times0.9} = 93$
2	6	0.99	$\frac{0.75\times10^7}{1.62\times10^5} = 46$	≈505	$\frac{505}{6\times0.99} = 85$
3	8	0.9	$\frac{0.75\times10^7}{1.33\times10^5} = 56.3$	≈520	$\frac{520}{8\times0.9} = 72$
4	8	0.99	$\frac{0.75\times10^7}{1.21\times10^5} = 62$	≈525	$\frac{525}{8\times0.99} = 66$

Table 8.7 Calculated result of self-cleaning time based on gravimetric concentration

No.	n/h^{-1}	η/% (arestance for $PM_{2.5}$)	N_0/N	$n\eta t$ in Fig. 6.1	$t=\frac{n\eta t}{n\eta}$/min
1	6	0.9	$\frac{0.75\times350}{6.3}=41.7$	≈500	93
2	6	0.99	$\frac{0.75\times350}{5.94}=44.2$	≈505	85
3	8	0.9	$\frac{0.75\times350}{4.8}=54.6$	≈515	72
4	8	0.9	$\frac{0.75\times700}{13}=40$	≈490	91

From the above calculation, for the house-hold air purifier with self-circulation mainly, the efficiency is usually larger than 0.9. In this case, the performance with the increased η is not so good as that with the increased n. For example, if $n=5$ h^{-1}, even though the efficiency is $\eta=1$, the value of $n\eta$ is 5 at most. While the efficiency is $\eta=0.9$ but the air change rate $n=6$ h^{-1}, the value of $n\eta$ is 5.4, which can reduce N by about 10%. Moreover, the increase of the corresponding pressure drop will be small. This should be paid special attention to by both designer and user.

It is shown from the above calculation that for the condition of introduction of outdoor air, the steady-state indoor particle concentration is larger than that with self-circulation condition by 1.3 times.

It should also be emphasized that in the equation the value of G_2 corresponds with $M=350$ μg/m^3. If the so-called non-guarantee rate is used, the value of M in Beijing is less than 350 μg/m^3 by more than 40%. If $M=700$ μg/m^3 is used, which is larger than 350 μg/m^3 by one times, N may be between 9 and 13 μg/m^3. For the former case, the period between replacements for air filter in the system is prolonged. While for the latter case, the period is shortened. So in terms of statistics, no matter the individual equipment or the system is concerned, it has no implication to calculate based on the non-guarantee rate based on the problem of particle pollution.

8.4.3 Calculation on Self-cleaning Time

The calculated self-cleaning time is shown in Table 8.7.

According to Table 1.2, 75 μg/m^3 is the upper limit of Class 2 standard, and 35 μg/m^3 is the upper limit of Class 1 standard.

It is the same as Table 8.6 in the previous section that the self-cleaning time for 10^7 pc/L is equivalent with that with the gravimetric concentration 350 μg/m^3.

It is ideal for the condition $n=6$ h^{-1} and $\eta=0.9$. Although the self-cleaning time is long, the indoor concentration N needed should not be as low as 6.3 μg/m^3. If the indoor concentration N should be larger than one times, the self-cleaning time needed will be much shorter.

8.5 Feasibility of Opening Windows

8.5.1 Outline

(1) When the air purifier is operated under the complete self-circulation condition, there is only infiltrated airflow indoors and no introduction of outdoor air. It is obvious that the volume of infiltrated airflow is too small to play a certain role as the outdoor air. So users will always ask a question: can air purifier be used under the condition of intermittent opening window? How large can be the opened area of the window? Or how long can be the opening period? The question from another aspect is: is it allowed to open a window gap during the continuous operation of air purifier? What is the impact on air purifier for these questions?

As mentioned before, during the severe pollution situation or worse condition, it is usually static wind. The value 0.35 m/s can be used as the calculated wind velocity in this book.

For the following calculations, some common conditions are set: room area 20 m^2, room height 2.5 m. There is one Class 4 window with size 1.5 m × 1.5 m. There are two sliding windows. The maximum window gas during the close of window is 0.57 mm. The maximum window gap during opening of window is 100 mm. For air purifier, $n = 6\,h^{-1}$ and the arrestance for $PM_{2.5}$ is 0.9. The ambient concentration is set 250 $\mu g/m^3$. The temperature differences between indoor and outdoor are set lower than 15 and 15–30 °C, respectively. The air densities are set 1.2 and 1.342 kg/m^3, respectively.

No matter there is infiltration air or inlet air through opened window, the calculation method for the combined effect of wind and thermal pressures in Chap. 7 should be adopted according to the above condition.

(2) According to the general expression for the combined effect of wind and thermal pressures in Chap. 7, when the temperature difference is less than 15 °C, the volume of the infiltration flow during the close of the window is:

$$\begin{aligned} Q'_{15} &= 0.7 \times \text{Gap length (m)} \times \text{Max. gap width (mm)} \\ &\quad \times \sqrt{\text{(Thermal pressure difference + Wind pressure difference)}} \\ &= 0.7 \times 7.5 \times 0.57 \times \sqrt{0.066 + 0.68} = 2.99 \times 0.86 = 2.58\,\text{m}^3/\text{h} \end{aligned}$$

The infiltrated particle during the close of window is equivalent to the corresponding indoor particle generation rate per volume (where the penetration coefficient is set 0.75):

$$G'_2 = \frac{0.75 \times \text{Volume of infiltration airflow} \times \text{Atmospheric dust concentration}}{60 \times \text{Room area} \times \text{Room height}}$$

$$= \frac{0.75 \times 2.58 \times 250}{60 \times 20 \times 2.5} = 0.16\,\mu\text{g/m}^3$$

The volume of inlet airflow during the opening of the window:

$$\begin{aligned} Q'_{15} &= 0.7 \times \text{Window height (m)} \times \text{Opening width (mm)} \\ &\quad \times \sqrt{(\text{Thermal pressure difference} + \text{Wind pressure difference})} \\ &= 0.7 \times 1.5 \times 100 \times \sqrt{0.066 + 0.68} = 105 \times 0.86 = 90.3\,\text{m}^3/\text{h} \end{aligned}$$

The infiltrated particles by opening of window can be considered as the corresponding indoor particle generation rate per unit volume (which is full penetration):

$$\begin{aligned} G''_2 &= \frac{\text{Volume of infiltration airflow} \times \text{Atmospheric dust concentration}}{60 \times \text{Room area} \times \text{Room height}} \\ &= \frac{90.3 \times 250}{60 \times 20 \times 2.5} = 7.6\,\mu\text{g/m}^3 \end{aligned}$$

(3) According to the general expression for the combined effect of wind and thermal pressures in Chap. 7, when the temperature difference is between 15 and 30 °C, the thermal pressure is increased from 0.68 to 1.35 Pa. The volume of the infiltration flow is:

$$Q'_{30} = 3.56\,\text{m}^3/\text{h}$$

$$G'_2 = 0.22\,\mu\text{g/(m}^3\cdot\text{min)}$$

$$Q''_{30} = 125\,\text{m}^3/\text{h}$$

$$G''_2 = 10.5\,\mu\text{g/(m}^3\cdot\text{min)}$$

8.5.2 *Continuously Opening Windows*

$$N = \psi \frac{60 G''_2}{n\eta}$$

(1) If the value of N is required to be less than 75 μg/m^3, when the temperature difference is less than 15 °C, n = 6 h^{-1}, and $\eta = 0.9$, the particle concentration per unit volume of the room can be calculated as follows:

$$G''_2 = \frac{\text{N} \times \text{n} \times \eta}{4 \times 60} = \frac{75 \times 6 \times 0.9}{1.9 \times 60} = 3.6\,\mu\text{g/m}^3\mu\text{g/(m}^3\cdot\text{min)}$$

Because 3.6 is about half of 7.6, it is similar as the situation for the window gap with width 10/2 = 5 cm. The corresponding flow rate of outdoor air is about 90.3/2 = 45 m^3/h. It can be used for two or three persons.

(2) When n = 8 h^{-1} and $\eta = 0.9$, $\psi = 1.7$. We obtain

$$G_2'' = 5.3\,\mu g/(m^3 \cdot min)$$

It is similar as the situation for the window gap with width 7 cm. The volume of the outdoor air is 70 m^3/h. It can be used for three or four persons.

(3) If N cannot exceed 75 μg/m^3 and the temperature difference is 15–30 °C, $G_2'' = 10.5\,\mu g/(m^3 \cdot min)$

It is known that when n = 6 h^{-1} and η = 0.9, the allowable value is $G_2'' = 3.6\,\mu g/(m^3 \cdot min)$. So the window gap with width 4 cm can be set. The volume of outdoor air is 50 m^3/h. It can be used for two or three persons.

It is known that when n = 8 h^{-1} and η = 0.9, the allowable value is $G_2'' = 5.3\,\mu g/(m^3 \cdot min)$. So the window gap with width 5 cm can be set. The volume of outdoor air is 63 m^3/h. It can be used for three or four persons.

(4) If the ambient concentration reaches 500 μg/m^3, when the temperature difference is less than 15 °C, the allowable window gap is the same as the situation (3). When the temperature difference is 15–30 °C, the allowable window gap is only a half of that in situation (3).

8.5.3 *Intermittently Opening Windows*

(1) After one night of operation, when the temperature difference is less than 15 °C, the indoor concentration in the morning will be stabilized in

$$N = 1.9\frac{60G'_2}{n\eta} = \frac{60 \times 0.16}{6 \times 0.9} = 3.4\,\mu g/m^3$$

(2) When the window with gap width 100 mm is open, we obtain

$$N = 1.9\frac{60G''_2}{n\eta} = 1.9\frac{60 \times 7.6}{5.4} = 160.4\,\mu g/m^3$$

So after the window is open, it takes the time t to reach stable value 160.4 μg/m^3.

$$\because \quad \frac{N_0}{N} = \frac{3.4}{160.4} = 0.021$$

$$\because\ 1-\frac{N_0}{N}=0.979$$

Look up Fig. 6.1, we obtain:

$$\mathrm{n\eta t}\approx 280,\ \mathrm{t}=\frac{280}{5.4}=51.9\,\mathrm{min}$$

The method to plot the pollution and cleaning processes has been given in author's book "Fundamentals of air cleaning technology" [2].

Concentration N_t at any time during the pollution process can be obtained from the following expression:

$$N_t\approx N-\Delta N\mathrm{e}^{-nt/60} \tag{8.7}$$

where N_t is the concentration at any time;
N is the steady-state pollution concentration;
n is the air change rate;
N_0 is the initial concentration;
ΔN is equal to $N-N_0$ (the pollution process).

The above equation can be expressed as:

$$N-N_t=(N-N_0)\mathrm{e}^{-nt/60}$$

$$\mathrm{e}^{-nt/60}=\frac{N-N_t}{N-N_0} \tag{8.8}$$

when nt = 30 $e^{\frac{-nt}{60}}=\frac{1}{e^{0.5}}=0.606$
nt = 60 $\mathrm{e}^{\frac{-nt}{60}}=\frac{1}{\mathrm{e}}=0.362$
nt = 120 $\mathrm{e}^{\frac{-nt}{60}}=\frac{1}{\mathrm{e}^2}=0.135$
nt = 180 $\mathrm{e}^{\frac{-nt}{60}}=\frac{1}{\mathrm{e}^3}=0.050$
nt = 240 $\mathrm{e}^{\frac{-nt}{60}}=\frac{1}{\mathrm{e}^4}=0.018$
nt = 300 $\mathrm{e}^{\frac{-nt}{60}}=\frac{1}{\mathrm{e}^5}=0.008.$

The pollution process can be seen from the rising curve shown in Fig. 8.7.

In this figure, the starting point is 3.4 μg/m^3, which is the initial concentration N_0. When the steady-state concentration $N=160.4$ μg/m^3 is reached, it takes the time $t=51.9$ min.

The process curve is plot from the end point towards the starting point. At time $t=50$ min, since $n=6\ \mathrm{h}^{-1}$, $nt=300$. So $e^{\frac{-nt}{60}}=0.008$. From Eq. (8.7), we can obtain:

$$N_t=160.4-(160.4-3.4)\times 0.008=160.4-157\times 0.008$$

This means the distance of 157×0.008 should be set downwards from the point linking the horizontal line with value 160.4 and the vertical line with value $t = 50$ min. At time $t=30$ min, the distance downwards should be 157×0.05. At time $t=20$ min, the distance downwards should be 157×0.135. When the window is open, the indoor concentration is required not to approach 160.4 μg/m³, instead only 100 μg/m³ is allowable. In this case, when $n=6\ \mathrm{h}^{-1}$, we obtain the following value from Eq. (8.10).

$$e^{0.1t} = \frac{160.4 - 3.4}{160.4 - 100} = \frac{157}{60.4} = 2.6$$

Then we obtain $t \approx 10$ min, which fits quite well with Fig. 8.7.

It is shown in Fig. 8.7 that after the window is open for 10 min and then closed, it takes very short period from 100 to 75 μg/m³, which is about 4 min. So it is feasible to open the window for 10 min.

(3) Self-cleaning with closed window

According to the previous calculation, when the window is closed, N can reach 3.4 μg/m³. So we obtain:

$$\frac{N_0}{N} = \frac{100}{3.4} = 30$$

$$\frac{N_0}{N} - 1 = 29$$

Look up Fig. 6.1, and we obtain $n\eta t \approx 480$. So $t=480/5.4=89$ min.

The cleaning process can be seen as the descending curve shown in Fig. 8.7.

Concentration N_t at any time during the cleaning process can be obtained from the following expression:

$$N_t \approx N - \Delta N e^{-nt/60} \tag{8.9}$$

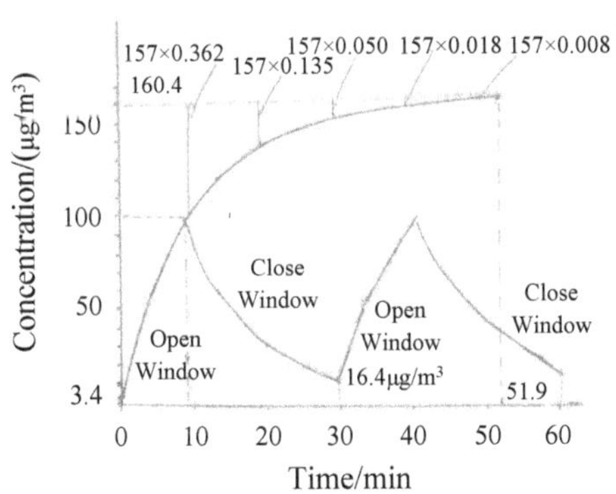

Fig. 8.7 Curves for pollution and cleaning processes

During the cleaning process, $\Delta N = N_0 - N$.

Where N_t is the concentration during the cleaning process;
N is the steady-state concentration reached during the cleaning process;
N_0 is the initial concentration.

So Eq. (8.9) can be converted into:

$$N - N_t = (N_0 - N)e^{-nt/60}$$

$$e^{-nt/60} = \frac{N - N_t}{N_{0-N}} \tag{8.10}$$

It is obvious that the indoor concentration is needed to be reduced to 3.4 μg/m³. Suppose the window is closed for 20 min, we obtain:

$$e^{-nt/60} = 0.135$$

Based on Eq. (8.9), the indoor concentration at this time will be

$$N_t = 3.4 + (100 - 3.4) \times 0.135 = 3.4 + 1.3 = 16.4\,\mu\text{g/m}^3$$

This is already very good level.
If the window is closed for 40 min, $e^{-nt/60} = 0.018$.

$$\therefore N_t \approx 3.4 + (100 - 3.4) \times 0.018 = 3.4 + 1.8 = 5.2\,\mu\text{g/m}^3$$

The above process can be referred to Fig. 8.7.
The above equations are used to calculate No. 1 curve in Fig. 6.4.
At $t = 11.5$ min, $nt = 60$ and the coefficient is 0.362. Then we obtain:

$$N_t = 25 + (590 - 25) \times 0.362 = 230\,\mu\text{g/m}^3$$

When the curve in Fig. 6.4 is checked, it is about 235 μg/m³.
At $t = 35$ min, $nt = 180$ and the coefficient is 0.05. Then we obtain:

$$N_t = 54\,\mu\text{g/m}^3$$

When the curve in Fig. 6.4 is checked, it is about 61 μg/m³.
It is shown that the calculation result is close to the experimental value.

(4) Ventilation with opened window

Suppose the window with width 100 mm is opened, the above procedures (2) and (3) are repeated.

(5) If the temperature difference between indoors and outdoors is 15–30 °C, the value of N for the situation (1) is increased by 1.375 times, which is 4.7 μg/m³; while the value of N for the situation (1) is increased by 1.382 times, which is 221.7 μg/m³.

The time needed for the indoor concentration to reach the allowable value 75 μg/m³ is:

$$e^{0.1t} = \frac{221.7 - 4.7}{221.7 - 75} = 1.483$$

$$\therefore t = 4\,\text{min}$$

The time needed for the indoor concentration to reach the allowable value 100 μg/m³ is:

$$e^{0.1t} = \frac{217.7}{121.7} = 1.789$$

$$\therefore t = 5.8 \approx 6\,\text{min}$$

These values are reduced by about 40% compared with the temperature difference 15–30 °C.

(6) If the ambient concentration is 500 μg/m³, both G'_2 and G''_2 are doubled. N will also be doubled. The corresponding size of the opened window should be reduced by a half. The condition 250 μg/m³ is still applicable.

(7) Summary

In order to maintain the indoor concentration at about 15 μg/m³, the self-cleaning time with the closed window should not be less than 20 min, and not be larger than 90 min. The time to ventilate with opened window should not exceed 10 min, and the size should not be larger than 100 min. If the actual air change rate is larger than 6 h^{-1} and the filtration efficiency is larger than 0.9, the condition for the opened window can be relaxed.

8.6 Applicable Area

8.6.1 Necessary Parameters

How large is the area of the room where the requirement can be satisfied with the application of air purifier? This is closely related to the flow rate it provides and the air cleanliness level.

It is known from the above calculation that under the condition of $n=6\ h^{-1}$ and $\eta=0.9$ in the severe pollution weather (350 $\mu g/m^3$), the indoor concentration N can be reduced to 16 and 6.3 $\mu g/m^3$ for the condition with outdoor air and the condition with indoor self-circulation, respectively.

Under the same condition, the larger the value of η is, the larger the self-cleaning time is. This is because the value of N reached will be smaller. But if the same value of N is reached, such as 75 $\mu g/m^3$, both of the self-cleaning times will be far less than 1 h.

Moreover, under the unfavorable condition with the temperature difference 30 °C, the window with the gap width 4 cm is still applicable.

So during the application of air purifier, the indoor air change rate should reach $6\ h^{-1}$ and its purification efficiency should not be less than 0.9. If it is lower than these indexes, the air cleanliness level reachable may be low, or the service area is too small.

8.6.2 Calculation on Applicable Area

If the room height is supposed 2.6 m, the applicable area should be:

$$F = \frac{Q_0}{6 \times \frac{0.9}{\eta'} \times H} \tag{8.11}$$

where, Q_0 is the nominal flow rate of air purifier, m^3/h;
η' is the filtration efficiency of air purifier;
H is the actual room height, m;

$$\because \quad \eta' = \frac{k}{n'} \tag{8.12}$$

where, k is the measured attenuation coefficient in the test chamber;
n' is $\frac{Q_0}{60\times30}$, which is equivalent with the air change rate per min in the test chamber;

If $k=0.09$ and $n'=0.1$, $\eta'=0.9$. When $H=2.6$, $Q_0=180\ m^3/h$ (it is obvious too small), we obtain:

$$F = \frac{Q_0}{6 \times 2.6} = \frac{180}{6 \times 2.6} = 11.5\,m^2$$

If $k=0.099$ and $n'=0.1$, $\eta'=0.99$. Other conditions are the same. Then we obtain:

$$F = \frac{180}{6 \times \frac{0.9}{0.99} \times 2.6} = 12.8\,m^2$$

If $k = 0.18$ and $n' = 0.2$, $\eta' = 0.9$. When $H = 2.6$, $Q_0 = 360\ \mathrm{m^3/h}$, then we obtain:

$$\mathrm{F} = \frac{360}{6 \times \frac{0.9}{0.9} \times 2.6} = 23.07\,\mathrm{m}^2$$

From the above analysis, if it is pursued that the air purifier should be as small as possible while the air change rate (nominal flow rate) is omitted, this is inappropriate for both manufacturer and user.

References

1. Z. Xu, G. Cao, Y. Zhang, L. Liang, Q. Zhou, Engineering application related issues of air cleaners—Discussion on several problems about air cleaners (4). J. HV&AC **47**(8), 14–17 (2017)
2. Z. Xu, *Fundamentals of Air Cleaning Technology*, 4th edn. (Science Press, Beijing, 2014)
3. H. Fukuyama, Airborne matter-nature and role. J. Soc. Heating, Air-Conditioning Sanit. Eng. Japan **49**(8), 57–60 (1975)
4. Y. Tu, G. Tu, X. Zhang, Study on particulate matter (PM2.5) filtration efficiency of ventilating air filters. J. HV&AC **46**(5), 49–54 (2016)
5. S. Kato, Selection method of air cleaning devices for different applications. Japan J. Air Cond. Heating, Refrig. News **11**(9), 103–100 (1971)
6. W. Zhang, J. Xie, J. Liu, Y. Diao, H. Yoshino, Research on minimum ventilation rate and air change rate of residential buildings at home and abroad. J. HV&AC **46**(10), 86–91 (2016)

Chapter 9
Test and Assessment of Air Purifier [1–2]

9.1 In Test Chamber

9.1.1 Chamber with Volume 30 m³

Since the air supply and exhaust openings are not uniform for air purifier (portable type, which is different from in-duct type) and their shapes are various, it is not suitable to be tested on the test rig. In 1985, Association of Home Appliance Manufacturers (AHAM) developed a test method for the above so-called portable air purifier. Air purifier is placed in the center of a sealed cabin with volume 30 m^3 (height × area = 2.5 m × 3.5 m × 3.4 m = 29.75 $m^3 \approx 30$ m^3). Tobacco smoke is used to obtain the initial concentration in the camber. Then the pollution removal efficiency of air purifier is measured for many times within a certain period continuously. Before the test, the smoke removal efficiency in self-decay process without air purifier is measured at first.

Figure 9.1 shows the schematic diagram of the test chamber.

Through multi-point measurement, the total attenuation coefficient can be calculated by the equation. When the natural attenuation coefficient is subtracted from it, the attenuation coefficient k after operation of air purifier can be obtained. Then Eq. (5.10) can be used to obtain the Clean Air Delivery Rate (CADR).

$$Q = 60\,kV$$

So the core of the entire measurement is to obtain the value of k.

Z. Xu, *Air Purifier*, https://doi.org/10.1007/978-981-13-2532-8_9

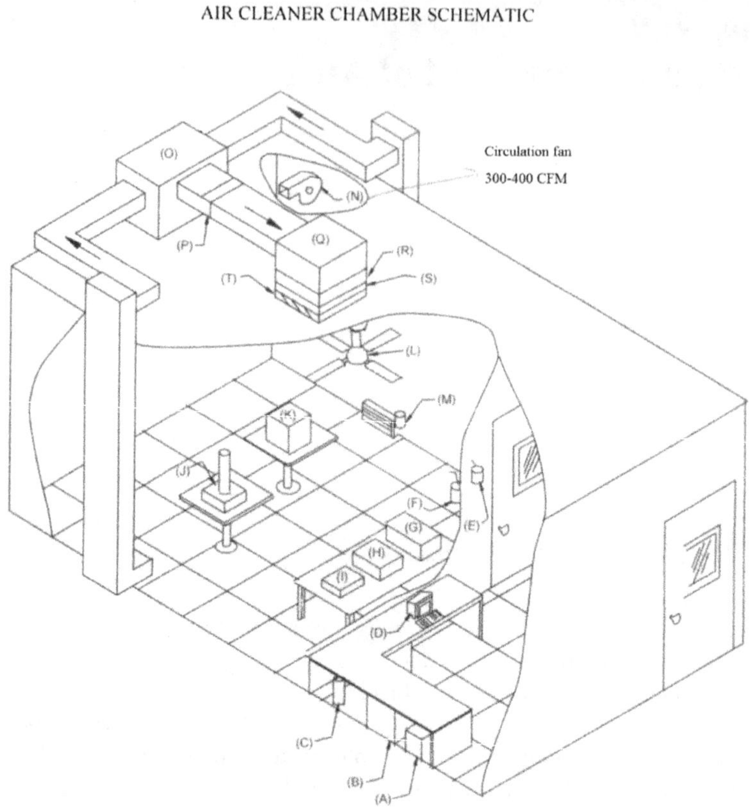

(A) Voltage stabilizer
(B) Data acquisition and control interface for measured device
(C) Air source (Air supply)
(D) Computer
(E) Inlet for soot
(F) Pollen generator
(G) Pollen monitor
(H) Dust monitor
(I) Smoke dilutor
(J) Dust generator

(K) Measured device
(L) Stirring fan
(M) Return air valve
(N) Circulation fan (300-400CFM)
(O) Humidifier
(P) Pre-filter
(Q) Fan section
(R) HEPA filter
(S) Electric heater
(T) Air supply valve

Fig. 9.1 Schematic diagram of the 30 m^3 test chamber in U.S. standard

9.1.2 *Natural Decay in Test Chamber*

9.1.2.1 Natural Decay Process

In both U.S. and Chinese national standards, it is required that the natural decay process in the test chamber should be performed before the performance of the air purifier is measured.

How much is the natural decay rate? Is it necessary to perform test? So the order of magnitude for the natural decay rate should be known.

In natural decay process, there are coagulation loss of particles, diffusional deposition loss on vertical surface and settlement loss on horizontal surface (room floor).

9.1.2.2 Coagulation Loss of Particles

With the relative movement of particles in the airflow (Brownian movement or that induced by airflow), large particles will be formed by mutual collision and attachment. This is termed as coagulation. With coagulation, the number of particles will be reduced.

There are two kinds of coagulation, which include the thermophoresis coagulation and turbulence coagulation.

(1) Thermophoresis coagulation

Thermophoresis coagulation is caused by Brownian movement of molecules, the particle concentration after time t will change from N_0 to N_t. It is expressed with Eq. (9.1), and the detailed derivation process for thermophoresis coagulation can be found in Refs. [3, 4].

$$N_t = \frac{N_0}{1 + N_0 k_0 t} \tag{9.1}$$

where, k_0 is the coagulation coefficient. k_0 is proportional to $\mathrm{D}d_\mathrm{p}$ (D is the diffusional coefficient).

Table 9.1 shows the value of k_0 under standard condition. Table 9.2 shows the variation of concentration during the thermophoresis coagulation process for monodisperse particles (about larger than 0.3 μm) when $k_0 = 5 \times 10^{-10}$ cm^3/s [4].

Table 9.1 Value of k_0 under standard condition

d_p/μm	k_0/($\times 10^{-10}$ cm^3/s)
0.01	67
0.1	8.6
1.0	3.5
10	3.0

Table 9.2 Time needed for concentration with reduction of 50% and size doubled by coagulation ($k_0 = 5 \times 10^{-10}$ cm^3/s)

Initial concentration N_0/(pc/cm^3)	Time needed when concentration becomes $0.5N_0$	Time needed when particle size is doubled ($N = 0.125N_0$)
10^{14}	20 μs	140 μs
10^{12}	2 ms	14 ms
10^{10}	0.2 s	1.4 s
10^{8}	20 s	140 s
10^{6}	33 min	4 h
10^{4}	55 h	16 d
10^{2}	231 d	4 × 365 d

Tobacco smoke is generated during test, which is required in U.S. standard on air purifier ANSI/AHAM AC-1-2006. According to the comprehensive analysis in literature, most of the tobacco smoke particles are within 0.06–0.5 μm, which is given in Fig. 9.2 [3].

According to the standard requirement for air purifier, the initial concentration after generation of tobacco smoke should be stabilized in 2×10^6–2×10^7 pc/L for particles with size ≥0.3 μm. Since the particle size distribution is not given in the standard, it is difficult to perform the calculation for the next step. When the general distribution of atmospheric dust is assumed, the concentration of particles with size ≥0.5 μm is about 1/3–1/2 of that with size ≥0.3 μm [3]. So the concentration for particles ≥0.5 μm can be set 10^7 pc/L. This means $N_0 = 10^7$ pc/L $= 10^4$ pc/cm^3 (this can also be set for particles ≥ 0.3 μm). Tobacco smoke particles are polydisperse. k_0 is used for monodisperse particles. It is much safer to set $k_0 = 5 \times 10^{-10}$ cm^3/s. The smaller the particle is, the easier the coagulation deposition will occur.

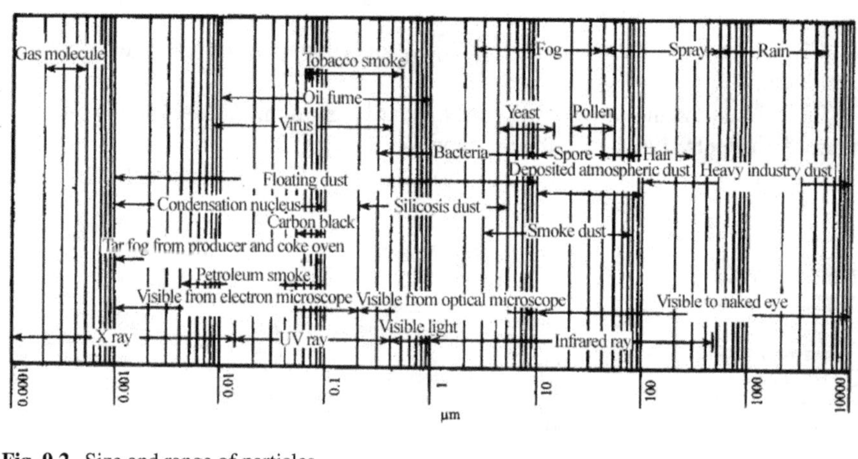

Fig. 9.2 Size and range of particles

It is required in the standard on air purifier that the test time should be $t = 1200$ s. From Eq. (9.1), we can obtain:

$$N_t = \frac{N_0}{1 + 10^4 \times 5 \times 10^{-10} \times 1200} = \frac{N_0}{1 + 0.6 \times 10^{-2}} = 0.994 N_0$$

This means that the concentration is only reduced by 6‰. The particle loss rate is 6‰. This conclusion has already been obtained during the early study of common aerosol. When $N_0 \leq 10^4$ pc/cm^3 (10^7pc/L), the thermophoresis coagulation can be ignored completely.

(2) Turbulence coagulation

(a) Turbulence coagulation is caused by turbulence flow, which is also called movement coagulation. Because it is too complex, the exact calculation method has not been given in the research field of aerosol. But it has been pointed out that the turbulence coagulation for particles with size 0.1 μm can be omitted. However, for particles with size larger than 10 μm, the turbulence coagulation is very important. For particles with size slightly larger than 1 μm, the effect of turbulence coagulation is much important than that of thermophoresis coagulation [4]. But the influence within 20 min is not very large. Suppose the turbulence coagulation of tobacco smoke particle is equivalent with the thermophoresis coagulation with particle size ≥ 0.3μm, the total coagulation loss is only 1.2%. Of course, this is only assumption. So the "self-purification" performance of test chamber will be limited (except the air velocity near the air supply outlet, the velocity induced by the air purifier in the room will not be larger than 0.3 m/s, so the effect of turbulence is very small).

(b) When the natural decay process is measured in the test chamber, in order to simulate the decay process during the operation of the unit, the air purifier is required to be turned off. And the circulation fan is turned on.

Based on U.S. standard, the flow rate of the circulation fan is 500–680 m^3/h (in Chinese standard, it is required to be 500–700 m^3/h). The volume of the airflow is not required to be adjustable. If the diameter of the air supply outlet is 20 cm, the velocity of the jet can reach 2–3.5 m/s, and the influencing range is very large. The volume of the test chamber is 30 m^3, $S = 3.5 \times 3.4 = 11.9$ m^2. The applicable area of air purifier is given in the equation by Chinese standard on air purifier:

$$S = 0.12Q$$

where, Q is the so-called CADR. According to the analysis in Sect. 5.2 of this book, $Q = \eta \cdot Q_0$. Q_0 is the nominal flow rate. When $\eta \geq 0.99$, $Q \approx Q_0$.

$$\therefore \quad Q_0 \approx Q = \frac{S}{0.12} = \frac{11.9}{0.12} = 99.2\,\text{m}^3/\text{h}$$

It is shown that the flow rate of the circulation fan is larger than that of the air purifier by several times under normal condition. The area of the air supply outlet in air purifier is similar as that of the circulation fan. So the air velocity for the latter case will be larger than the former by several times. The effect of the turbulence flow is much larger than that from the air purifier. In this case, there is a certain extent of the turbulence coagulation. The coagulation loss will be larger than the actual situation.

(3) Diffusional Deposition Loss on Vertical Surface

The inertial deposition on vertical surface can be completely omitted [3]. Only diffusional deposition is considered. With diffusional deposition loss, the initial concentration inside the test chamber will be reduced continuously. It will not be kept constant. So it is applicable to the diffusional deposition with convection without air supply.

The number of particles by diffusional deposition on vertical surface per unit area is:

$$N_s = \frac{V}{s}\left(1 - e^{-\frac{sDt}{V\delta}}\right) N_0 \ \text{pc/m}^2 \tag{9.2}$$

where, s is the area of the vertical surface, cm^2;
V is the space volume, cm^3;
D is the diffusional coefficient, cm^2/s; For particles with size 0.3 μm, it is about 1.1×10^{-6};
δ is the thickness of the molecular diffusional layer; Although it is difficult to be determined, it is about the magnitude of 20 μm according to the experiment [5];
t is the time, s.

Now let's take an example for calculation:
Suppose $V = 30\ \text{cm}^3 = 30 \times 10^3 \times 10^3\ \text{cm}^3 = 30 \times 10^6\ \text{cm}^3$.

$$s = 2 \times (3.4 + 3.5) \times 2.5 \times 10^4 = 34.5 \times 10^4\ \text{cm}^2$$
$$t = 1200\ \text{s}$$

This is the situation of the test chamber for air purifier.

Then the diffusional coefficient D of smoke particles is considered based on the particles with diameter 0.3 μm, from Eq. (9.2) we can obtain:

$$N_s = \frac{30 \times 10^6}{34.5 \times 10^4}\left[1 - e^{-34.5\times10^4\times1.1\times10^{-6}\times1.2\times10^3/(30\times10^6\times20\times10^{-4})}\right] \times 10^4$$
$$= 87 \times \left(1 - e^{-0.0076}\right) \times 10^4 = 6612\ \text{pc/cm}^2$$

The total diffusional deposition on four surfaces is $6612 \times 34.5 \times 10^4$ pc $= 228 \times 10^7$ pc

It is occupies $\frac{228\times10^7}{30\times10^6\times10^4} = 0.76 \times 10^{-2} = 0.76$ of the initial concentration.

(4) Settlement Deposition Loss on Horizontal Surface (Floor)

The correction equation for settlement deposition on horizontal surface is [3]:

$$N_g = \alpha V_s t N \tag{9.3}$$

where, V_s is the deposition velocity, cm/s; it is related to the area weighted diameter; For the standard particle size distribution for particle with size $\geq$0.5 μm in air, the area weighted diameter can be calculated to be $D_s = 0.98$ μm. So for particles with size $\geq$0.5 μm in air, the deposition rate can be considered to be about 1 μm with the deposition velocity $V_s = 0.006$ cm/s.
α is the correction factor when various deposition factors are considered. For the large horizontal surface 11.9 m^2 in the test chamber, it is 1 for $D_s \approx 1$ μm.
t is the deposition time, s; which is set 1200 s;
N is the initial concentration, pc/cm^3; which is set 10^4 pc/cm^3;

Then we obtain:

$$N_g = 1 \times 0.006\,\text{cm/s} \times 1200\,\text{s} \times 10^4\,\text{pc/cm}^3 = 7.18 \times 10^4\,\text{pc/cm}^2$$

The total number of particles deposited onto the floor within 20 min is $11.9 \times 10^4 \times 7.18 \times 10^4 = 0.85 \times 10^{10}$ pc

The loss rate for particles with size larger than 0.5 μm is:

$$\frac{0.85 \times 10^{10}}{30 \times 10^6 \times 10^4} = 0.028 = 2.8\%$$

The particle size distribution is unknown in the test chamber where tobacco smoke is generated. All the deposition can be considered as the particles with size 0.5 μm. Then $V_s = 0.001$ cm/s and α should be set 1.3. $N_g = 1.56 \times 10^4$ pc/cm^2. The particle loss rate is reduced to 0.0061, or 0.61%.

The equation of this deposition rate has been proved correct by both the experimental result and the relationship between the airborne bacteria and the deposition bacteria from U.S. NASA standard [3].

(5) Sum of Natural Decay Loss

The sum of the particle loss by coagulation, diffusion and deposition in the test chamber is about:

$$0.012 + 0.0076 + 0.0061 = 0.026 = 2.6\%$$

This loss rate is insignificant compared with the efficiency of air purifier which is more than 0.9.

In the test chamber, there should be a stable natural decay process (similar as the background value). It is unnecessary to perform this kind of test each time. Since particle loss rate of this kind is very small, the result is the same which is applied to each measured equipment. So there is no influence when it is omitted. According to the measurement data on many types of equipment, the natural decay rate is about 3%. If the air supply velocity is reduced, the decay rate will be much smaller. According to the test on Cang-Qiong KJ12-CFZ electrostatic air purifier by Shanghai Municipal Bureau of Quality and Technical Supervision, the natural decay rate within 30 min in the test chamber is 2% [6].

Therefore, it is suggested in this book that it is unnecessary to perform the natural decay test for air purifier in the test chamber. The background concentration can be replaced at most. The circulation fan can be cancelled. Small fan with adjustable flow rate can be replaced near the sample prototype. Or according to the requirement for the blank control test in GB21551.3-2010, the component responsible for removing aerosol inside the air purifier should be removed, and then the flow rate of the fan is adjusted to the original nominal value.

9.2 Details During Measurement

9.2.1 Positioning Locations

The measured air purifier should not be placed at the center of the test chamber or the small room (shown in Fig. 9.3). This does not conform to the actual application situation. Usually air purifier is placed near the wall, near the furniture, or at the corner of the room. If it is placed at the center of the room (shown in Fig. 9.4), air will be supplied mainly from the front/top, or from the face inclined to the front/top, or from the side/front/top, or from three faces. Air will be returned from the lower front face, or from the side or from the bottom. It is obvious that there lacks clean air to dilute at the back of the air purifier. And there is vortex. Even though air will be supplied from the top side as the flower pot structure, air purifier will usually be placed near the wall or at the corner. For the ceiling cassette fan-coil unit, air is supplied downwards or from the four edges. It is even more difficult to understand why test is performed at the height of 700 mm.

So it is recommended to test the air purifier on the floor center in the back of test chamber or on the table (it is determined by its actual application, and the ceiling type or the wall mounted is not included). The front face is 0.5 m from the wall. It is shown in Fig. 9.4.

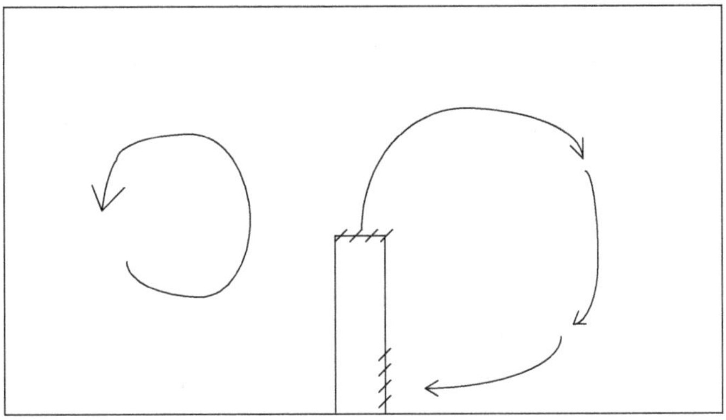

Fig. 9.3 Air purifier placed in the center of the test chamber

Fig. 9.4 Positioning of sampling points

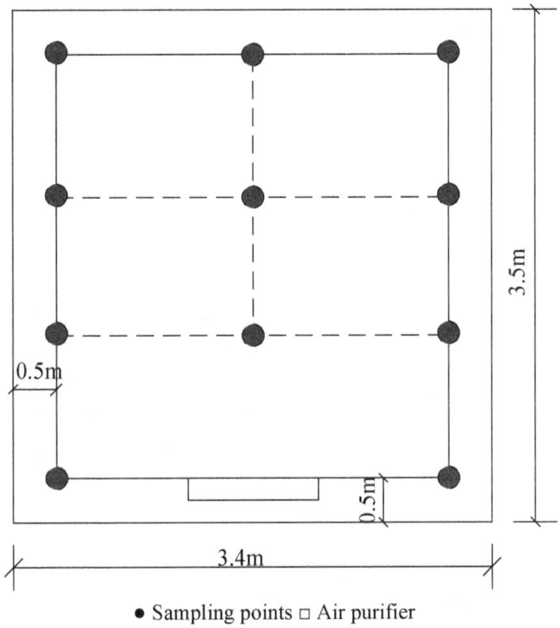

9.2.2 *Sampling Points*

The number and location of measuring points should be clearly defined. Otherwise, there must be a difference between the results. For example, if it is not defined in the standard, when more sampling points are placed in the front space as shown in Fig. 9.2, favorable results will be obtained. So if there is no uniform provisions for the layout of the measuring points, there will be poor comparability between results, and fairness will be lost.

It is recommended to uniformly place the sampling points at the breathing height 0.8 m under sitting state in the horizontal plane which is 0.5 m from the each wall according to the concept of the activity region. As shown in Fig. 8.3, the minimum number of sampling points can be eleven (Or more sampling points can be specified. But uniform provisions must be made. It should not be specified in principle, such as "not less than 9 points". This is because the measurement results need to be compared). The location of the sampling points must be specified. Label can be made on the floor. If there is specification beforehand, the positioning of objects in the chamber can be avoided. Even though there is a slight hindrance, since it is the same, measurement results can be comparable.

9.2.3 Method

It should be clearly stated that whether the particle counting method or the gravimetric concentration method is used. For the former situation, the sampling flow rate should also be defined.

9.2.4 Reading Data

(1) By convention, all the sampling values should be recorded for use in the next step. It should not be specified that the value less than certain limit (such as 50 pc/L) or the sampling point should be deleted. If sampling points are within the mainstream of the clean supplied airflow and the flow rate of the air purifier is large, the reading value may be low. Since average value is used, it is unfair by doing so. It is specified in Chinese national standard GB50591-2010 "Code for construction and acceptance of cleanroom" that "for the single non-random abnormal reading caused by measurement error or abnormally low particle concentration (air is reported clean), when it will influence the calculation result, this abnormal reading is allowed to be deleted. But it should be noted in the original record" and "only one reading is allowed to be deleted for each measurement space". It can be referred for measurement of air purifier. Of course, if there is more than one reading for such situation, the measurement can be specified invalid, which should be started again.

(2) According to Eq. (5.8), we know

$$\frac{N_t}{N_0} = \mathrm{e}^{-kt}$$

It is known that the value of N_t during each test should be the average concentration on the plane of the room. It is not the steady-state concentration. The concentration varies each time. So the concentration at the same time cannot be represented by the readings point by point. In the national standard on air purifier, it is not specified whether samplings are performed at the same time or point by point. From the reading each 2 min, it belongs to the latter case. So the concentration at one sampling position each time cannot represent the room average concentration at this time. So they cannot be used to calculate the value of k.

(3) In order to obtain the simultaneous readings, the on-line particle monitoring system should be adopted. Several probes should be placed at each sampling positions. Remote real-time monitoring, alarm setting, data recording, report compilation and print output should be performed through computers in the control room, which are connected to the particle counters.

9.3 Calculation on *k*

9.3.1 Calculation Graph

Equation (5.8) can be rewritten as:

$$\frac{N_0}{N_t} = \mathrm{e}^{\mathrm{kt}} \tag{9.3}$$

$$\mathrm{kt} = \ln \frac{N_0}{N_t} \tag{9.4}$$

Figure 9.5 can be plotted.

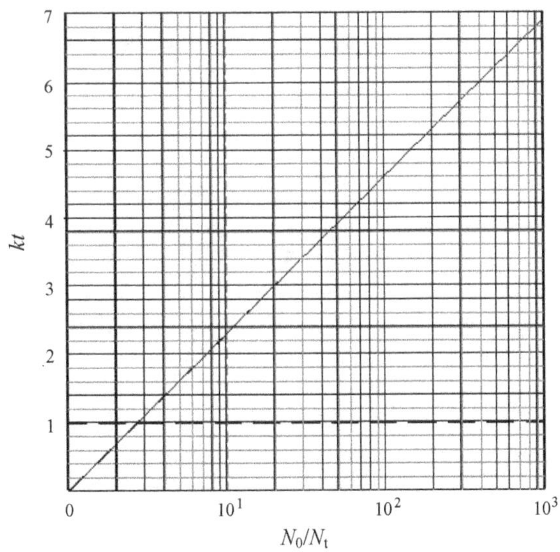

Fig. 9.5 Calculation graph on attenuation coefficient in the test chamber

9.3.2 Examples

Suppose the value of N_0/N_t is measured after 20 min, where N_t is the average value of various sampling points at the time 20 min. Look at Fig. 9.5, the value of kt can be obtained. Then $k = kt/20$.

It is assumed $N_0/N_t = 10$ after 20 min. From Eq. (9.1) or Fig. 9.5, we know

$$\ln 10 = 2.3 = \mathrm{kt}$$

$$\mathrm{k} = \frac{2.3}{20} = 0.115$$

When the equivalent air change rate in the test chamber is $n' = 0.12\ \mathrm{min}^{-1}$, $\eta = k/n' = 0.115/0.12 = 0.96$.

If after 20 min it becomes $N_0/N_t = 6$, from the figure we know $kt = 1.8$.

$$\therefore\quad \mathrm{k} = \frac{1.8}{20} = 0.09$$
$$\eta = \frac{0.09}{0.12} = 0.75$$

If after 30 min it becomes $N_0/N_t = 10$, we know

$$\mathrm{k} = \frac{2.3}{30} = 0.077$$
$$\eta = \frac{k}{\mathrm{n}'} = \frac{0.077}{0.12} = 0.64$$

In Chap. 5, calculation shows that all the η is 0.9. $n_b = 2n_a$. $n_a = 0.1\ \mathrm{min}^{-1}$. The purification ability of air purifier b can be six times as the ability of air purifier a at the time 20 min during the measurement. The above result can also be proved from Fig. 9.5.

$$\because\quad k = n'\eta,\ \text{then}\ \mathrm{k_a} = 0.1 \times 0.9 = 0.09$$
$$\mathrm{k_b} = 0.2 \times 0.9 = 0.18$$
$$\therefore\quad \mathrm{tk_a} = 1.8,\ \mathrm{tk_b} = 3.6$$

Look up Fig. 9.5, we know when $\mathrm{tk_a} = 1.8$, it is $\frac{N_0}{N_{at}} = 6$; when $\mathrm{tk_b} = 3.6$, it is $\frac{N_0}{N_{bt}} = 36$.

Then N_{bt}: $N_{at} = 1/6$. This means that after 20 min during experiment, the ability of air purifier b: the ability of air purifier a = 6:1.

The larger the time t needed to reach stable is, the larger the difference between N_t will be. The multiplication times of the corresponding purification capability is much larger.

If $t = 40$ min, $tk_a = 3.6$ and $tk_b = 7.2$.

Look up Fig. 9.5, we know: $\frac{N_0}{N_{at}} = 36$ and $\frac{N_0}{N_{bt}} = 1000$. This means that the difference of the air purification capability between them at time 40 min reaches 30 times. In this case, the value of N_{bt} is already very small.

9.3.3 Attention

It should be noted that Fig. 9.5 is only suitable for the test chamber where there is only initial concentration but not generation of particles. It is different from Fig. 6.1. It cannot be used for calculation for the actual application situation with indoor particle generation. The value of *k* includes the value of natural decay.

9.4 Assessment

9.4.1 Existing Problems

From the previous discussion, the parameter CADR cannot be used to precisely predict the purification capability of air purifier under various conditions. It is inappropriate to be used for assessment and comparison between air purifiers.

From the above analysis, it is no need to correlate the attenuation coefficient k and the designed CADR. The parameter *k* itself can be used to reflect the purification capability of air purifier. As CADR, it is not scientific to judge "good" or "bad" just based on *k* only. For an air purifier with size larger by one time and flow rate larger by one time then the one shown in Fig. 9.6, the value of *k* is definitely larger than the smaller air purifier. It only means that the value of *k* is large, but it cannot be used to explain that the performance of one air purifier is superior to the other. So the method with the test chamber and the value of *k* obtained in the test chamber may be misleading.

In order to objectively compare the purification performance of air purifier, the value of *k* must be specified for the unit air volume. For a good air filter in air purifier, the pressure drop is small and the flow rate is large. The size of the air filter is not increased. If the flow rate for another air purifier needs to be increased, the size of the air filter should also be increased, so that the pressure drop can be maintained the same. So the volume of the latter is also increased. Although the values of *k* per unit air volume are the same, it is obvious that the purification performance of the air purifier with smaller volume and smaller air filter is much better. Therefore, the key element of unit volume should also be emphasized. In order to save energy, the factor considering the unit energy consumption should be included.

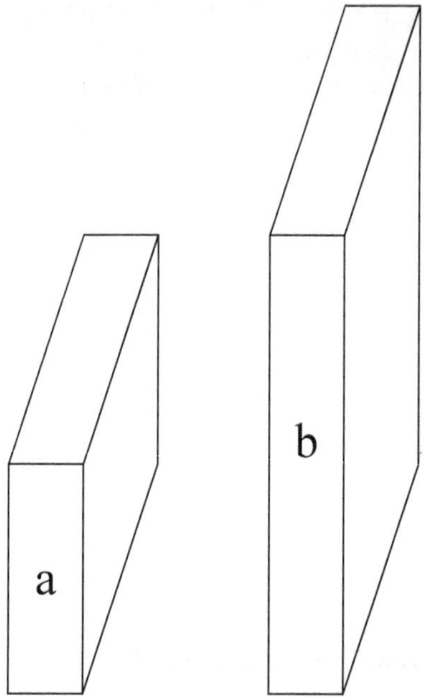

Fig. 9.6 Two air purifiers with difference of filtration performance by one time

9.4.2 Comprehensive Index

The comprehensive index to express the air purification performance per unit energy consumption and the energy efficiency is called the Cleaning Performance under Energy Efficiency (CPEE). It can be expressed as:

$$CPEE = \frac{k}{20QV_aP} \tag{9.5}$$

where Q is the flow rate to test the air purifier, which should be nominal flow rate, m^3/min; Since the test period is 20 min, it is $20Q$;
V_a is the volume of the external rectangle connected to the air purifier, m^3;
P is the energy consumption, kW;

The larger the CPEE value is, the better it is.

For example, when $k = 0.12$, $Q = 300$ m^3/h, $V_a = 0.6 \times 0.4 \times 0.2 = 0.048$ m^3, $P = 0.1$ kW, then we obtain CPEE = 0.25.

When $k = 0.12$, $Q = 340$ m^3/h, $V_a = 0.048$ m^3, $P = 0.11$ kW, then we obtain CPEE = 0.2.

The volumes of two air purifiers are the same. The values of k are equal. But the flow rate of the second air purifier is larger than that of the first one. This means the efficiency of the second air purifier may be lower. Although the same value of k can be kept, the performance of the second air purifier is poorer than that of the first one, and CPEE for the second air purifier is smaller than the first one.

Obviously the evaluation method of this kind is reasonable. CPEE is only one comprehensive index, which only reflects the comprehensive capability of air purifier (purification efficiency and energy consumption). It cannot reflect the quantitative proportional relationship for the purification capability of the comprehensive performance. If only the strong purification capability is required while other factors are not considered, the one with large value of k can be selected.

During the application of the above comprehensive index, it can be specified that the measured CPEE cannot be lower than the nominal value by a certain percentage, such as 10%.

References

1. Z. Xu, X. Feng, Y. Zhang, L. Liang, Q. Zhou, Natural attenuation of air cleaners in environmental test chamber and its influencing factors—discussion on several problems about air cleaners (3). J. HV&AC **47**(8), 11–13 (2017)
2. Z. Xu, G. Cao, Y. Zhang, W. Niu, H. Pan, Performance testing method of air cleaners in environmental test chamber-discussion on several problems about air cleaners (5). J. HV&AC **47**(8), 18–19 (2017)
3. Z. Xu, *Fundamentals of Air Cleaning Technology*, 4th edn. (Science Press, Beijing, 2014)
4. W. C. Hinds, *Aerosol Technology: Properties, Behavior, and Measurement of Airborne Particles*, 2nd edn. (Wiley-Interscience, 1999)
5. N.A. Fuchs, *The Mechanics of Aerosols* (Dover Publications, New York, 1989)
6. H. Mao, *Study on Improvement of Indoor Air Quality with Electrostatic Air Cleaner* (Tongji University, 2008)

CPSIA information can be obtained
at www.ICGtesting.com
Printed in the USA
LVHW04*0037180918
590499LV00006B/41/P

9 789811 325311